T0231828

Randomized Response and Indirect Questioning Techniques in Surveys

STATISTICS: Textbooks and Monographs

Recent Titles

Multivariate Statistical Analysis, Second Edition, Revised and Expanded, *Narayan C. Giri*

Computational Methods in Statistics and Econometrics, *Hisashi Tanizaki*

Applied Sequential Methodologies: Real-World Examples with Data Analysis, *edited by Nitis Mukhopadhyay, Sujay Datta, and Saibal Chattopadhyay*

Handbook of Beta Distribution and Its Applications, *edited by Arjun K. Gupta and Saralees Nadarajah*

Item Response Theory: Parameter Estimation Techniques, Second Edition, *edited by Frank B. Baker and Seock-Ho Kim*

Statistical Methods in Computer Security, *edited by William W. S. Chen*

Elementary Statistical Quality Control, Second Edition, *John T. Burr*

Data Analysis of Asymmetric Structures, *Takayuki Saito and Hiroshi Yadohisa*

Mathematical Statistics with Applications, *Asha Seth Kapadia, Wenyaw Chan, and Lemuel Moyé*

Advances on Models, Characterizations and Applications, *N. Balakrishnan, I. G. Bairamov, and O. L. Gebizlioglu*

Survey Sampling: Theory and Methods, Second Edition, *Arijit Chaudhuri and Horst Stenger*

Statistical Design of Experiments with Engineering Applications, *Kamel Rekab and Muzaffar Shaikh*

Quality by Experimental Design, Third Edition, *Thomas B. Barker*

Handbook of Parallel Computing and Statistics, *Erricos John Kontoghiorghes*

Statistical Inference Based on Divergence Measures, *Leandro Pardo*

A Kalman Filter Primer, *Randy Eubank*

Introductory Statistical Inference, *Nitis Mukhopadhyay*

Handbook of Statistical Distributions with Applications, *K. Krishnamoorthy*

A Course on Queueing Models, *Joti Lal Jain, Sri Gopal Mohanty, and Walter Böhm*

Univariate and Multivariate General Linear Models: Theory and Applications with SAS, Second Edition, *Kevin Kim and Neil Timm*

Randomization Tests, Fourth Edition, *Eugene S. Edgington and Patrick Onghena*

Design and Analysis of Experiments: Classical and Regression Approaches with SAS, *Leonard C. Onyiah*

Analytical Methods for Risk Management: A Systems Engineering Perspective, *Paul R. Garvey*

Confidence Intervals in Generalized Regression Models, *Esa Uusipaikka*

Introduction to Spatial Econometrics, *James LeSage and R. Kelley Pace*

Acceptance Sampling in Quality Control, *Edward G. Schilling and Dean V. Neubauer*

Applied Statistical Inference with MINITAB®, *Sally A. Lesik*

Nonparametric Statistical Inference, Fifth Edition, *Jean Dickinson Gibbons and Subhabrata Chakraborti*

Bayesian Model Selection and Statistical Modeling, *Tomohiro Ando*

Handbook of Empirical Economics and Finance, *Aman Ullah and David E. A. Giles*

Randomized Response and Indirect Questioning Techniques in Surveys, *Arijit Chaudhuri*

Randomized Response and Indirect Questioning Techniques in Surveys

Arijit Chaudhuri

Indian Statistical Institute

Calcutta, India

CRC Press
Taylor & Francis Group
Boca Raton   London   New York

CRC Press is an imprint of the
Taylor & Francis Group, an **informa** business

A CHAPMAN & HALL BOOK

Chapman & Hall/CRC
Taylor & Francis Group
6000 Broken Sound Parkway NW, Suite 300
Boca Raton, FL 33487-2742

First issued in paperback 2017

© 2011 by Taylor and Francis Group, LLC
Chapman & Hall/CRC is an imprint of Taylor & Francis Group, an Informa business

No claim to original U.S. Government works

ISBN-13: 978-1-4398-3657-6 (hbk)
ISBN-13: 978-1-138-11542-2 (pbk)

Visit the Taylor & Francis Web site at
http://www.taylorandfrancis.com

and the CRC Press Web site at
http://www.crcpress.com

To Bulu

Contents

Preface

On "Randomized Response" there exist two widely cited and circulated books. I happen to be a coauthor of one of them. They are (1) *Randomized Response: A Method for Sensitive Surveys* (1986) by J.A. Fox and P.E. Tracy and (2) *Randomized Response: Theory and Techniques* (1988) by Arijit Chaudhuri and Rahul Mukerjee.

Nevertheless, it seems that one more book is required. The subject of randomized response (RR) is growing phenomenally in theory but not keeping pace in practice, as desired. Our purpose here is to take stock of the former and to renew a plea for the latter.

The subject of RR is an aspect of survey sampling. A finite survey population of a known number of identifiable people, N, is to be addressed. The object is to estimate the population total Y of a variable y which is defined on it. In case y denotes an attribute bearing only a 1/0 value, then instead of the total, the mean $\overline{Y} = Y/N$, which is the same as the proportion, say, θ of the people bearing a specified characteristic A in the community of N people that needs to be estimated. For this, a sample s from the population $U = (1, \ldots, i, \ldots, N)$, denoting the labels of the identified people, has to be chosen and surveyed.

A new and vital problem arises if the subject of study refers to a sensitive and, even more, to a stigmatizing rather than an innocuous feature of a person which has to be addressed when sampled. It is often hard to realize a direct response (DR) from a sampled person by dint of a direct query. For example, one may like to assess the percentage of people in a community who are typically among habitual law breakers, alcoholics, gamblers, tax evaders, oppressors of spouses, or in collusion with terrorists or extremists nationally or internationally, or drunken drivers, and many others. An investigator may be hesitant to directly ask the sampled person whether the person bears any of these characteristics in order to obtain a "Yes" or "No" response. Even if an investigator gathers enough courage to do so, a likely consequence is either a false response or a refusal to answer. Therefore, it is deemed rather impossible to derive an unbiased estimate of the parameter θ that is of one's interest. In order to overcome this problem to the extent possible, Warner (1965) proposed a novel technique of implementing a survey by trying to gather a suitable RR from each sampled person. On specifying a rightly designed scheme of sampling with the RRs gathered, an unbiased estimator for θ is then available, as shown by Warner (1965). Moreover, a measure of accuracy in estimation is also available in terms of the standard error and the estimated coefficient of variation of the resulting estimator. This development rapidly generated an ever-growing literature on the emerging alternative techniques for eliciting suitable RRs from people by discussing their pros and cons. In Chapter 1 of this treatise we set forth why this particular book is needed. The subsequent chapters describe and review in brief some aspects of the developments achieved so far on RR. The emphasis will naturally be on those areas with active involvement by the author.

Acknowledgments

I am indebted to the Director of the Indian Statistical Institute and my colleagues in its Applied Statistics Unit for their valuable time in helping me complete this book.

1

Why We Need One More Monograph on Randomized Response

An overwhelming message revealed in a major portion of the articles on randomized response (RR) and randomized response techniques (RRTs) seems to be the following: A simple random sample (SRS) with replacement (WR) may be drawn, from a population and from a person sampled, irrespective of how often the person was sampled, an RR may be truthfully procured from the person by dint of a suitably addressed RR device. Since for an SRSWR every draw independently produces a respondent with an equal probability, the theory on RR developed so far turns out to be extremely simple.

Almost invariably, the RR theory in the literature till date is apparently connected to the SRSWR and the population proportion bearing a sensitive attribute, being essentially a finite population mean, is unbiasedly estimated by the arithmetic mean of the suitably transformed values of all the RRs gathered in respect of the persons sampled no matter how many times. The variance of this simple mean and an unbiased estimator of the variance of this estimator are easily derived. Comparative performances of the alternative estimators with the varying RR schemes are also studied in great detail with little difficulty.

In this book, we first recall that in case of DR surveys, other than the overall sample mean based on an SRSWR, there exist alternative unbiased estimators for the population mean taking into account the set of distinct units rather persons who happen to be sampled in the predetermined number of draws for the sample. Also, for simple inverse sampling, an SRSWR is taken in successive draws continued until a predetermined number of distinct persons appear in the sample. Since the publication of several books and monographs on RR, some remarkable developments have taken place to examine how the subject modifies itself in case of RR rather than DR being generated in these two special ways in SRSWR. We briefly but critically intend to take into account these emerging developments. It is somewhat more interesting to again observe that RRTs may be implemented no matter how samples of potential respondents are chosen. If the sample is chosen suitably, the RRs generated from the sampled persons in appropriate ways may be put to analysis in many prudent ways to yield estimators for the percentage of people with the sensitive features in a specific community amenable to comparative studies. Since various alternative estimation procedures are available for contention, they do not relegate to the familiar ones when the SRSWR, in particular, is employed as a special case of the sampling scheme.

In Chaudhuri and Mukerjee's (1988) book, Rao-Blackwellization was shown to yield an improved method of estimation in case some respondents offered to divulge DRs, given an option. Here the respondent-wise data actually gathered are available for the DRs and separately for the RRs. Currently, a theory is being developed to permit the respondent an option to give a DR rather than an RR without divulging to the investigator the actual option so exercised.

How to measure the protection of privacy afforded by an RRT actually employed has been studied in case of SRSWR. How to extend the theory to the situations employing unequal probability sampling, even without replacement, is currently being studied and a theoretically promising literature is growing.

Besides RRTs, alternative techniques for indirect questioning in data generation covering sensitive issues are also described in the literature. A brief review of these is also promised in this treatise.

The later chapters of this book will address the following issues:

> *Chapter 2.* Warner (1965) in his pioneering model to tackle the problem of gathering trustworthy data relating to dichotomous sensitive information on a human feature A and its complement A^c, essentially recommended approaching a person to be sampled in a community, with a box containing a certain number of indistinguishable cards of which a proportion $p(0 < p \neq 1/2 < 1)$ is marked A and the rest marked A^c. The person on request is required to draw a "random" card and respond by answering "Yes" for a "match" between the card type and the person's own real characteristic or a "No" for a "nonmatch" before returning the card to the box. No matter how a sample is chosen it is possible to derive, for every person labeled i, which is 1, ..., N with a value y_i as 1 if i bears the characteristic A, or o if the characteristic is A^c, an unbiased estimator r_i instead of y_i, with a variance in terms of p, with respect to the randomization experiment involved in data gathering. Warner restricted exclusively to samples chosen by SRSWR taking an RR by the above method independently every time a person is chosen randomly. Here we briefly consider a current development with variations in estimation procedures, taking account of the units that are distinct using only one RR, from the person or all the RRs separately generated every time the person appears in the sample, keeping the total number of draws in the sample intact or the number of draws are allowed to continue till a predetermined number of distinct units appear in the sample. This theoretical exercise is motivated because in the context of DR surveys, Basu (1958), Pathak (1962), Des Raj and Khamis (1958), Lanke (1975a), Chikkagoudar (1966), Asok (1980) among others, developed detailed theories of estimation with these three aspects of SRSWR.

Our collective interest is about how these are modified with DR being replaced by RR.

Chapter 3. In the first six chapters in the book by Chaudhuri and Mukerjee (1988) all the RRTs described were stated to apply to people sampled exclusively following the method of SRS and SRSWR. Consequently the theory of estimation developed was restricted to SRSWR situation alone. More importantly, the simple means of suitable transforms of all the RRs generated, no matter how many times the same person gives out the RRs independently on every appearance in the sample, formed the basis of estimation of the proportion of people bearing a sensitive attribute and also the means of quantitative characteristics bearing social stigma, formed the basis of estimation of all related parameters. In that publication, a finite survey population set up was given due attention for the first time in Chapter 7. In this updated book, we emphasize that application of RRTs in the estimation of attribute or quantitative features is valid for selection of respondents in a general manner. The only requisite in the estimation is the positive inclusion probability of every individual and also of every pair of distinct members of the community intended to be covered. In this chapter, we illustrate how this may be accomplished for several RRTs developed to date and is also of our interest.

Chapter 4. Warner's (1965) RR scheme intended to estimate the unknown proportion θ of people in a community of N persons bearing a sensitive attribute A, is applied to generate "Yes," or "No" responses from persons chosen independently by making n draws by SRS method. Assuming λ as the unknown probability of getting a "Yes" response, the probability of getting $n_1 (0 \leq n_1 \leq n)$ "Yes" responses, according to the "binomial law," is

$$P(n_1) = \binom{n}{n_1} \lambda^{n_1} (1 - \lambda)^{n-n_1}, \quad 0 < \lambda < 1.$$

But

$$\lambda = Prob(\text{a "Yes" response}) = p\theta + (1 - p)(1 - \theta)$$
$$= (1 - p) + \theta(2p - 1), \quad 0 \leq \theta \leq 1;$$

as a matter of fact, λ cannot have a value beyond the range

$$(1 - p, p) \quad \text{if } p > \frac{1}{2}$$

$$\text{or} \quad (p, 1 - p) \quad \text{if } p < \frac{1}{2};$$

of course, $p \neq 1/2$.

Yet, $\hat{\lambda} = n_1/n$ is both an "unbiased" and the "maximum likelihood estimator" (MLE) of λ. Hence,

$$\hat{\theta} = \frac{\hat{\lambda} - (1-p)}{2p-1} = \frac{\frac{n_1}{n} - (1-p)}{(2p-1)}$$

is "unbiased" for θ according to Warner (1965) who mistakenly also described this $\hat{\theta}$ as the MLE of θ as well.

Singh (1976, 1978) pointed out this fallacy by observing that the space,

$$\left[-\frac{(1-p)}{2p-1}, \frac{p}{2p-1} \right]$$

of possible values of $\hat{\theta}$ does not coincide with that of θ which is [0, 1]. As a matter of fact, the MLE of θ under Warner's RR scheme based on SRSWR is

$$\begin{aligned} \theta^* &= \hat{\theta} \quad \text{if } 0 \le \hat{\theta} \le 1, \\ &= 0 \quad \text{if } \hat{\theta} < 0, \\ &= 1 \quad \text{if } \hat{\theta} > 1. \end{aligned}$$

But fresh data are needed to treat the MLEs for θ under various RR devices when samples may be taken in various ways and not necessarily by SRSWR method.

Chapter 5. The RR theory so far discussed, may be called CRR, that is, "Compulsory Randomized Response" theory. This is because the investigator anticipates the feature of interest to enquire about carries a social stigma and so decides to adopt an RR device to elicit reliable answers to certain queries of relevance.

But it is possible to offer each respondent an option by either of the two following ways to respond in alternate but truthful ways.

The respondent may not judge the issue in question sensitive enough and may offer to give out the truth about the genuine characteristics enquired about.

As another alternative the respondent may be given an option either to give out the truth or to respond following the RR device offered with a choice not to disclose which of the two options is actually exercised.

Both of these two alternatives are recognized as ORR, that is, Optional Randomized Response devices, as elaborated in the literature.

Chapter 6. If a person to be sampled really perceives the feature in question to be stigmatizing, as is believed by the investigator, a cooperation in giving out the truthful response as per an RR device offered is expected, provided the RR is trusted not to reveal the true sensitive feature A, say, of the respondent. So, how far the respondent's privacy is protected by the RR device put in operation is a very crucial question. In case of a "Yes/No" or "Either/Or" response about A or A^c in the contrasting situation, a rational respondent is expected to weigh the chance to be detected as bearing the stigmatizing characteristic A, say, if the RR device executed demands a "Yes" rather than a "No" response. If $\theta(0 < \theta < 1)$ is the proportion bearing A in the community and if SRSWR is the sample-selection method employed, then on every draw of a sample respondent the probability of the person bearing A remains θ. But this is not the case for selection without replacement, especially if the population size N is a moderate number. This probability varies perceptibly over successive draws for a sample size not too small relative to N. The theory developed covering RRTs tied to SRSWR so far is rather easy to comprehend. But as our stand in this book is to attach real importance to RRs gathered from execution of general sample selection schemes, the developments concerning the latter case will be stressed in what follows. A rather comprehensive ground seems to have been covered so far, especially after a couple of publications have appeared on the scene.

Chapter 7. RR techniques are available to cover quantitative variables including those related to stigmatizing characteristics. As possible examples, one may cite under-reported amounts of income in tax returns, costs of treatments on AIDS, number of induced abortions, amounts gained and lost in gambling, number of days of drunken driving last quarter of the year and others. Writing y_i as the value of such a variable y for the individual labeled i in the population $U = (1 \dots, i, \dots, N)$ the problem addressed is to estimate $Y = \sum_1^N y_i$ on choosing a sample s with a probability $p(s)$ according to a suitable design p and gathering RRs as z_is for i in s on adopting a convenient RR device. The requisite task is to bring forth a serviceable estimator, derive a variance expression for it along with a suitable estimator for it, and explore optimal choices, if practicable. Optional RR and protection of privacy are associated aspects to be studied as well to the extent possible, availing ourselves of the available literature. As usual in this book we avoid citing the cases in the literature treated by our colleagues inspiring our enthusiasm to the extent possible for an adequate coverage.

Chapter 8. It is an undeniable fact that DR techniques alone are not capable of providing reliably serviceable procedures for estimation of parameters relating to sensitive items of interest. Though RRTs are developing in prolific manners both intensively and extensively they are far from popular in public applications. Sheer theoretical richness is not really enough for the viability of an idea that seems to be the case as a matter of fact. There are two main criticisms against RRT, (1) it demands a little too much of an active participation from a respondent who must be clever and eager enough to appreciate the objectives of the survey, and (2) in addition, be credible about the personal privacy being well-protected despite the spontaneous revelation through the medium of the RRT, accepted for execution. Besides, there is no provision in an RR device to have any clarifying dialogue in course of its implementation, between the respondent and the investigator. As a consequence, there have been moves to discard the practice of RR technologies. To our knowledge three techniques have thus far emerged called Item Count Technique (ITC), Nominative Technique, and the Three-Card Method, as possible better alternatives. Instead of presenting a comprehensive review of these three alternative procedures to tackle sensitive, rather stigmatizing issues by reliably relevant data gathering and putting them to analytical uses, we shall attempt here at communicating our own views towards their development.

Chapter 9. Research on RR theory and techniques as discussed so far through the published literature mainly revolves around (1) prescribing more and more accurate estimators for a population proportion by developing sharper data-gathering devices, (2) assessing the extent of privacy protection afforded by respective procedures, each for respondents chosen by SRSWR, and (3) covering the "Mean or Total" estimation of a quantitative variable that is stigmatizing through general sampling schemes as well, possibly in optimal manners. The present contributor's thesis has thus far revealed endeavors to clarify that "an RR device as distinct from another" is to be applied on a person no matter how selected and suitable unbiased or moderately biased estimators are to be employed based on the RRs gathered in suitable manners. Measures of privacy protection RR-device-wise are laid down and estimates of means and totals based on RRs through individuals by various schemes of sampling are recommended to be examined with respect to their measures of accuracies corresponding to divergent levels at which the privacies are protected by the RR devices yielding the databases.

From now on, it behooves us to pay attention to the interesting things in the context of RRs and indirect questioning techniques, and resulting data processing issues, mostly the others, especially those the contemporary researchers are propounding over the years.

Chapter 10. Chapter 10 recounts certain simulation results.

A few of the relevant references related to this chapter are listed below.

1. Chaudhuri, A. 2001a
2. Chaudhuri, A. and Saha, A. 2005
3. Chaudhuri, A. and Pal, S. 2008
4. Chaudhuri, A., Bose, M., and Dihidar, K. 2009a
5. Chaudhuri, A., Bose, M., and Dihidar, K. 2009b
6. Chaudhuri, A. and Christofides, T.C. 2007
7. Chaudhuri, A. and Christofides, T.C. 2008
8. Chaudhuri, A., Christofides, T.C., and Saha, A. 2009
9. Warner, 1965
10. Basu, 1958
11. Pathak, 1962
12. Des Raj and Khamis, 1958
13. Lanke, 1975a
14. Chikkagoudar, 1966
15. Asok, 1980
16. Chaudhuri, A. and Mukerjee, R., 1988
17. Singh, 1976
18. Singh, 1978

2

Warner's Randomized Response Technique

Introduction

Every person in a given community is supposed to bear a stigmatizing feature, say, A or its complement A^c, which may or may not be stigmatizing. For example, one may be a tax-evader, a characteristic A worth hiding or may be an honest tax payer bearing A^c, a feature not needed to be concealed. Again a person may be disloyal or loyal to his/her boss who is a corporate body general manager. Either characteristic seems worth being kept a secret. It may be socially useful to form an accurate idea about the percentage in a given community of people that may bear a socially disapproved feature, say, A. Fearing that it may be pragmatically unethical for a social researcher to directly ask a person about bearing A or not bearing it and also speculating that a truthful answer or any answer at all may not be elicited by direct questioning about A, it appeared for long a social necessity to hit upon an appropriate trick to successfully deal with such a situation. In 1965, Warner just provided one as published in the *Journal of American Statistical Association*.

Warner Model

A box containing a large number of cards of the same shape, thickness, length, breadth, weight, and color but differing in proportion, $p(0 < p \neq (1/2) < 1)$ of them marked A and the others marked A^c is presented to a person to be sampled in the community. The person is requested to shake the box, draw one of the cards, note its mark unseen by the enquirer and return it to the box after saying "Yes" if the card type drawn "matches" the person's real characteristic, A or A^c or "No" if "there is no match." With every person we associate a 1/0 variable y which is y_i for the ith person in the community of N persons such that

$$y_i = 1 \quad \text{if } i\text{th person bears } A$$
$$= 0 \quad \text{if the person bears } A^c.$$

The mission is to appropriately estimate the proportion $\theta = Y/N, (Y = \Sigma_1^N y_i)$ in the community actually bearing A.

Let $I_i = 1$ if Warner's above-noted randomized response (RR) experiment yields a "Match" in the card type and the ith person's y_i- value
$= 0$, otherwise.

We shall generically write E_R, V_R, the expectation, variance operators with respect to Warner's or any other RR technique to be covered in this book. For Warner's RR technique, it follows that

$$E_R(I_i) = py_i + (1-p)(1-y_i)$$
$$= (1-p) + (2p-1)y_i,$$

leading to $r_i = [I_i - (1-p)]/(2p-1)$ and $E_R(r_i) = y_i$.

Again, since $I_i^2 = I_i$ and $y_i^2 = y_i$,

$$V_R(I_i) = E_R(I_i)(1 - E_R(I_i))$$
$$= p(1-p), \quad \text{and hence}$$
$$V_i = V_R(r_i) = \frac{p(1-p)}{(2p-1)^2} = \phi_w.$$

Here r_i are independent variables across i in $U = (1, ..., i, ..., N)$.

Simple Random Sampling with Replacement

Whatever might have been the reason, Warner (1965) gave his theory on randomized response rechnique (RRT) only in respect of the selection of individuals by simple random sampling with replacement (SRSWR). Also, everytime an individual happens to be selected, an RR is independently gathered on request from the individual and an unbiased estimator for θ is derived using all these RRs.

Writing N for the population-size, y_K, I_K as the values y_i, I_i for the unit chosen on the Kth draw $(K = 1, ..., n)$, n as the total number of draws in the sample,

$$r_K = \frac{I_K - (1-p)}{(2p-1)}, \quad K = 1, ..., n$$

has the expectation $E_R(r_K) = y_K$ and $\bar{r} = (1/n)\sum_{k=1}^{n} r_k$ has the expectation $E_R(\bar{r}) = (1/n)\sum_{k=1}^{n} y_K = \bar{y}$. Also, $V_R(\bar{r}) = (\phi_w/n)$.

Writing E_p, V_p as the expectation–variance operators generically for any sampling design p and $E = E_p E_R = E_R E_p$ and $V = E_p V_R + V_p E_R = E_R V_p + V_R E_p$, one has

$$E(\bar{r}) = E_p(\bar{y}) = \frac{1}{N}\sum_{1}^{N} y_i = \frac{Y}{N} = \bar{Y} = \theta$$

in this case and

$$V(\bar{r}) = E_p\left(\frac{\phi_w}{n}\right) + V_p(\bar{y}) = \frac{\phi_w}{n} + \frac{\sigma^2}{n},$$

writing $\sigma^2 = (1/N)\sum_{1}^{N}(y_i - \bar{Y})^2$ and since $y_i^2 = y_i$, Warner got $V(\bar{r}) = 1/n$ $[\phi_w + \theta(1 - \theta)]$.

Writing $s_r^2 = 1/(n-1)\sum_{k=1}^{n}(r_K - \bar{r})^2$, it follows that $v = (1/n)s_r^2$ is an unbiased estimator for $V(\bar{r})$. Warner (1965), however, observed $\sum_{k=1}^{n} I_K$ to follow the binomial distribution $b(.; n, \lambda)$, writing $\lambda = p\theta + (1 - p)(1 - \theta) = (1 - p) + (2p - 1)\theta$ because in case of SRSWR, this λ is the probability that a person, whenever drawn, may truthfully answer "Yes" on applying Warner's RRT. Noting $n_1 = \sum_{k=1}^{n} I_K$ as the observed total number of "Yes" responses in n draws by SRSWR and $\hat{\lambda} = n_1/n$, Warner (1965) worked out

$$v' = \frac{\hat{\lambda}(1 - \hat{\lambda})}{(n - 1)(2p - 1)^2}$$

as an unbiased estimator for $v(\bar{r})$. It may be easily checked that $v = v'$.

Note: It is possible that r_i may take a value outside the interval [0,1] and consequently, $\bar{r}$ as well may have a value less than 0 or greater than 1. In such cases, it is impossible to interpret $\bar{r}$ as serviceable though it is an unbiased estimator for θ such that $0 \le \theta \le 1$. This unavoidable companion of RRTs has remained an insurmountable problem.

Chaudhuri and Pal's Estimators

In case of the availability of DR as the values y_k for $k = 1, \ldots, n$ from an SRSWR in n draws it is of course well-known that $\bar{y} = (1/n)\sum_{k=1}^{n} y_k$ is an unbiased estimator for $\bar{Y} = (1/N)\sum_{1}^{N} y_i$, $V_p(\bar{y}) = \sigma^2/n$, and $1/(n-1)\sum_{k=1}^{n}(y_k - \bar{y})^2$ is an unbiased estimator for $V_p(\bar{y})$. But it is also well known that $\bar{y}$ is not a good estimator, because if $m(1 \leq m \leq n)$ is the number of distinct units found in SRSWR in n draws, then as Basu (1958, 1969) and Basu and Ghosh (1967) have shown, the "set of these m distinct units" along with their associated y values constitutes a "sufficient," rather "the minimal sufficient" statistic corresponding to the detailed statistic $((k,y_k)\,|\,k=1, \dots, n)$ from the SRSWR in n draws; it follows by the principle of Rao-Blackwellization that corresponding to this sample s, $\bar{y}' =$ the mean of y_i's for i in the set s_m of m distinct units in s, is also unbiased for $\bar{Y}$ and more importantly $V_p(\bar{y}') \leq V_p(\bar{y})$.

This raised the question about a possible comparison, in the context of RR rather than DR, between $\bar{r} = (1/n)\sum_{k=1}^{n} r_k$ and the altered $\bar{r}' = (1/m)\sum_{i \in s_m} r_i = \bar{r}(m)$.

Another question is if any other comparable unbiased estimator for $\bar{Y}$ may also be devised based on y_i for i in s_m in case of DR and correspondingly r_i for i in s_m in case of RR-based surveys.

From Chaudhuri and Pal (2008), we may reproduce the essentials in brief. Let m distinct units in s_m out of s chosen in n draws by SRSWR be each approached and requested to give only one RR on employing Warner's RRT as r_i for i in s_m; then, the estimator for θ they propose is $\bar{r}' = (1/m)\sum_{i \in s_m} r_i$. As shown by Basu (1958), we know that

$$E_p\left(\frac{1}{m}\sum_{i \in s_m} y_i\right) = \bar{Y},$$

and hence likewise

$$E_p\left(\frac{1}{m}\sum_{i \in s_m} r_i\right) = \frac{\sum r_i}{N} = \bar{R},$$

and

$$E(\bar{r}') = E_R\left(\frac{1}{N}\sum_1^N r_i\right) = \bar{Y} = \theta,$$

and likewise

$$E(\bar{r}') = E_p E_R\left(\frac{1}{m}\sum_{i \in s_m} r_i\right) = E_p\left(\frac{1}{m}\sum_{i \in s_m} y_i\right) = \bar{Y} = \theta.$$

Using Pathak's (1962) DR-related findings, one gets

$$V(\bar{r}') = E_p V_R(\bar{r}') + V_p E_R(\bar{r}') = E_p\left[\frac{\phi_w}{m}\right] + \left[E_p\left(\frac{1}{m}\right) - \frac{1}{N}\right]S^2,$$

where

$$S^2 = \frac{1}{N-1}\sum_{1}^{N}(y_i - \bar{Y})^2 = \frac{N\theta(1-\theta)}{N-1}.$$

This is also shown by Mangat et al. (1995).
So,

$$V(\bar{r}') = \phi_w E_p\left(\frac{1}{m}\right) + \left(E_p\left(\frac{1}{m}\right) - \frac{1}{N}\right)\frac{N\theta(1-\theta)}{N-1}.$$

To derive an unbiased estimator for $V(\bar{r}')$, we borrow DR-based results from Pathak (1962).
Let

$$v_r(m) = \frac{1}{(m-1)}\sum_{i\in s_m}(r_i - \bar{r}')^2$$

and

$$v(m) = \frac{1}{(m-1)}\sum_{i\in s_m}(y_i - \bar{y}')^2.$$

Further, let

$$v_{2r}(\bar{r}') = v_r(m)\left[\frac{1^{n-1} + 2^{n-1} + \cdots + (N-1)^{n-1}}{N^n}\left(\frac{N}{N-1}\right)\frac{C_m(n) - C_m(n-1)}{C_m(n)}\right],$$

$$= v_r(m)C_2.$$

Here,

$$C_m(n) = \sum_{i=0}^{m-1}(-1)\binom{m}{i}(m-i)^n, \quad i = 1, 2, \ldots, (n-1)$$

and

$$C_2 = \frac{1^{n-1} + 2^{n-1} + \cdots + (N-1)^{n-1}}{N^n}\left(\frac{N}{N-1}\right)\frac{C_m(n) - C_m(n-1)}{C_m(n)}.$$

Then,

$$E(v_{2r}(\bar{r}')) = C_2 E_p E_R[v_r(m)]$$

$$= C_2 E_p E_R \left[\frac{m-1}{m} \sum_{i \in s_m} r_i^2 - \frac{1}{m} \sum_{i \neq} \sum_{i \in s_m} r_i r_j \right] \Big/ (m-1)$$

$$= C_2 E_p \left[\frac{p(1-p)}{(2p-1)^2} + \frac{1}{m} \sum_{i \in s_m} y_i^2 - \frac{1}{m(m-1)} \sum_{i \neq} \sum_{j \in s_w} y_i y_j \right]$$

$$= C_2 [\phi_w + E_p[v(m)]]$$

Pathak (1962) had given

$$C_2 E_p(v(m)) = V(\bar{y}') = \left[E_p \left(\frac{1}{m} \right) - \frac{1}{N} \right] S^2.$$

So, $v_{2r}(\bar{r}')$ is an unbiased estimator of

$$V(\bar{y}') = E_p \left(\frac{1}{m} - \frac{1}{N} \right) \frac{N\theta(1-\theta)}{(N-1)}.$$

So, an unbiased estimator for $V(\bar{r}')$ is $v(\bar{r}') = v_{2r}(\bar{r}') + (1/m - C_2) - \phi_w = v_2$, using Pathak (1962) one may further consider the following:

$$C_3 = \frac{C_{(m-1)}(n-1)}{C_m(n)}, \ v_{3r}(\bar{r}(m)) = C_3 v_r(m)$$

$$C_4 = \left[\left(\frac{1}{m} - \frac{1}{N} \right) + \frac{N-1}{N^n - N} \right], \ v_{4r}(\bar{r}(m)) = C_4 v_r(m)$$

$$C_5 = \left[\left(\frac{1}{m} - \frac{1}{N} \right) + N^{1-n} \left(1 - \frac{1}{m} \right) \right], \ v_{5r}(\bar{r}(m)) = C_5 v_r(m)$$

So, for $V(\bar{y}')$ the following unbiased estimators emerge, namely,

$$v_j = v_{jr}(\bar{r}') + \left(\frac{1}{m} - C_j \right) \phi_w, \ j = 2,\ldots,5.$$

Further writing

$$V(\bar{y}') = E_R V_p(\bar{r}') + V_R E_p(\bar{r}')$$

$$= E_R E_p[v_p(\bar{r}')] + \frac{\phi_w}{N},$$

with $v_p(\bar{r}')$ as an unbiased estimator for $V_p(\bar{r}')$ such that $E_p v_p(\bar{r}') = V_p(\bar{r}')$, we get the following four more unbiased estimators for $V(\bar{y}')$ utilizing throughout the corresponding DR-based results of Pathak (1962), namely

$$v_j = v_{jr}(\bar{r}') + \frac{\phi_w}{N}, j = 2,\ldots,5.$$

Chaudhuri and Pal (2008) further considered corresponding to the DR-based Horvitz and Thompson's (1952) estimator $t_H = (1/N)\sum_{i\in s_m} y_i/\pi_i$, its RR-version $\bar{r}'' = (1/N)\sum_{i\in s_m} r_i/\pi_i$. Here $\pi_i = \sum_{s\ni i} p(s)$ is the well-known inclusion probability of a unit i for a design p with $p(s)$ as the probability of choosing a sample s. Chaudhuri and Stenger (2005) pointed out that $\pi_i > 0 \forall i$ if Y is to admit an unbiased estimator.

For SRSWR in n draws

$$\pi_i = 1 - \left(\frac{N-1}{N}\right)^n, \quad \pi_{ij} = 1 - 2\left(\frac{N-1}{N}\right)^n + \left(\frac{N-2}{N}\right)^n.$$

Also,

$$E_p(m) = \sum_{1}^{N} \pi_i = N\left(1 - \left(\frac{N-1}{N}\right)^n\right).$$

Writing $I_{si} = 1$ if $i \in s$, and 0, else, $I_{sij} = 1/0$, if $i, j \in s$, else

$$E_p(I_{si}) = \sum_{s\ni i} p(s) = \pi_i, \quad E_p(I_{sij}) = \sum_{s\ni i,j} p(s) = \pi_{ij}$$

$$\sum_{1}^{N} \pi_i = \sum_{s} \gamma(s)p(s),$$

if $\gamma(s) =$ number of distinct units in s.

$$\sum_{j\neq i} \pi_{ij} = \sum_{s\ni i} \gamma(s)p(s) - \pi_i,$$

$$\sum_{i\neq}\sum_{j} \pi_{ij} = \sum_{s} \gamma^2(s)p(s) - \sum_{}\pi_i = V_p(\gamma(s)) + \gamma(\gamma - 1),$$

when $\gamma = \sum_s \gamma(s)p(s)$, which is the expected number of distinct units realized in a sample s. So, for a DR-survey if one employs

$$\bar{y}'' = \frac{1}{N}\sum_{i\in s} \frac{y_i}{\pi_i} = \frac{1}{N}\sum_{1}^{N} \frac{y_i}{\pi_i} I_{si}$$

one gets

$$V_p(\bar{y}'') = \frac{1}{N^2}\left[\sum \frac{y_i^2}{\pi_i^2}V_p(I_{si}) + \sum_{i\neq}\sum_j \frac{y_i}{\pi_i}\frac{y_j}{\pi_j}Cov_p(I_{si},I_{sj})\right],$$

writing Cov_p as the covariance operator for a design p.

$$= \frac{1}{N^2}\left[\sum_i y_i^2 \frac{1-\pi_i}{\pi_i} + \sum_{i\neq}\sum_j y_i y_j \frac{\pi_{ij}-\pi_i\pi_j}{\pi_i\pi_j}\right]$$

(because $V_p(I_{si}) = \pi_i(1-\pi_i), Cov_p(I_{si},I_{sj}) = \pi_{ij}-\pi_i\pi_j) = V_{HT}(\bar{y}''))$.

An alternative formula for $V_p(\bar{y}'')$ is available from Chaudhuri and Pal (2002) as

$$V_p(\bar{y}'') = \frac{1}{N^2}\left[\sum_i^N \sum_j^N \left(\frac{y_i}{\pi_i}-\frac{y_j}{\pi_j}\right)^2 (\pi_i\pi_j-\pi_{ij}) + \sum \frac{y_i^2}{\pi_i}\beta_i\right]$$

$$= V_{CP}(\bar{y}'');$$

here,

$$\beta_i = 1 + \frac{1}{\pi_i}\sum_{j\neq i}\pi_{ij} - \gamma.$$

If $\gamma(s) = \gamma$ for every s with $p(s) > 0 \forall s$, then $\beta_i = 0 \forall i(= 1,\ldots,N)$, and one gets

$$V_p(\bar{y}'') = \sum_i^N \sum_j^N \left(\frac{y_i}{\pi_i}-\frac{y_j}{\pi_j}\right)^2 (\pi_i\pi_j-\pi_{ij}) = V_{YG}(\bar{y}'')$$

available only for a fixed-effective sample-size design as given by Yates and Grundy (1953). The number of distinct units in a sample is called the effective size of a sample. So, the following alternative unbiased estimators followed for $V_p(\bar{y}'')$, namely

$$v_{HT}(\bar{y}'') = \frac{1}{N^2}\left[\sum_i y_i^2 \frac{(1-\pi_i)}{\pi_i}\frac{I_{si}}{\pi_i} + \sum_{i\neq}\sum_j y_i y_j \frac{(\pi_{ij}-\pi_i\pi_j)}{\pi_i\pi_j}\frac{I_{sij}}{\pi_{ij}}\right]$$

and

$$v_{CP}(\bar{y}'') = \frac{1}{N^2}\left[\sum_i \sum_{<j}\left(\frac{y_i}{\pi_i}-\frac{y_j}{\pi_j}\right)^2 (\pi_i\pi_j-\pi_{ij})\frac{I_{sij}}{\pi_{ij}} + \sum_i \frac{y_i^2}{\pi_i}\beta_i\frac{I_{si}}{\pi_i}\right]$$

which reduces to

$$v_{YG}(\overline{y}'') = \frac{1}{N^2}\left[\sum_i\sum_{<j}\left(\frac{y_i}{\pi_i} - \frac{y_j}{\pi_j}\right)^2(\pi_i\pi_j - \pi_{ij})\frac{I_{sij}}{\pi_{ij}}\right]$$

in case of a "fixed-effective sample-size" design, provided in each case $\pi_{ij} > 0\,\forall i \neq j$, which we assume to be the case for every sampling design we may employ.

We shall give a common notation $v_p(\overline{y}'')$ for each $v_{HT}(.)$, $v_{CP}(.)$, and $v_{YG}(.)$ above. Now $V(\overline{r}'') = E_pV_R(\overline{r}'') + V_pE_R(\overline{r}'')$

$$= \left(\frac{1}{N^2}\phi_w\right)\sum_1^N\frac{1}{\pi_i} + V_p(\overline{y}'').$$

Let $\underline{Y} = (y_1,\ldots,y_i,\ldots,y_N), \underline{R} = (r_1,\ldots,r_i,\ldots,r_N)$ and $V_p(\overline{r}'') = V_p(\overline{y}'')\big|_{\underline{Y}=\underline{R}}$ which is but the same as $V_p(\overline{y}'')$ evaluated at $\underline{Y} = \underline{R}$. We may then denote the common $v_p(\overline{y}'')\big|_{\underline{Y}=\underline{R}}$ by $v_p(\overline{r}'')$, that is any of the three formulae $v_{HT}(\overline{y}''), v_{CP}(\overline{y}''), v_{YG}(\overline{y}'')$ each evaluated at $\underline{Y} = \underline{R}$. We may now observe that

$$E_pE_Rv_{HT}(\overline{r}'') = \frac{1}{N^2}\left[\sum_i\frac{1-\pi_i}{\pi_i}\left(\sum_1^N y_i^2 + \phi_w\right) + \sum_{i\neq}\sum_j\left(\frac{\pi_{ij}-\pi_i\pi_j}{\pi_i\pi_j}\right)y_iy_j\right]$$

$$= \frac{1}{N^2}\phi_w\sum_i\left(\frac{1-\pi_i}{\pi_i}\right) + V_{HT}(\overline{y}'')$$

and

$$V_{HT}(\overline{r}'') = \frac{\phi_w}{N^2}\sum\frac{1}{\pi_i} + V_{HT}(\overline{y}'').$$

So,

$$v_1(\overline{r}'') = v_{HT}(\overline{r}'') + \frac{1}{N^2}\phi_w\sum\frac{I_{si}}{\pi_i}$$

is an unbiased estimator for $V_{HT}(\overline{r}'')$, because $V_{HT}(\overline{r}'') = E_pV_R(\overline{r}'') + V_pE_R(\overline{r}'')$, that is

$$V_{HT}(\overline{r}'') = E_p\left[\frac{1}{N^2}\phi_w\sum\frac{1}{\pi_i^2}I_{si}\right] + V_{HT}(\overline{y}'') = \frac{1}{N^2}\phi_w\sum\frac{1}{\pi_i} + V_{HT}(\overline{y}'')$$

Also, we have

$$V_{HT}(\overline{r}'') = E_R V_{HT}(\overline{r}'') + V_R E_p(\overline{r}'')$$

$$= \frac{1}{N^2}\left[\sum \frac{1-\pi_i}{\pi_i}(y_i^2 + \phi_w) + \sum_{i\neq}\sum_j y_i y_j \frac{\pi_{ij} - \pi_i\pi_j}{\pi_i\pi_j}\right] + \frac{\phi_w}{N}$$

$$= V_{HT}(\overline{y}'') + \frac{1}{N^2}\phi_w \sum \frac{1}{\pi_i}$$

$$= E_R E_p v_{HT}(\overline{r}'') + \frac{\phi_w}{N}.$$

Consequently, $v_2(\overline{r}'') = v_{HT}(\overline{r}'') + \phi_w/N$ is another unbiased estimator for $V(\overline{r}'')$.

Next, we see

$$V_{CP}(\overline{r}'') = E_p V_R(\overline{r}'') + V_p E_R(\overline{r}'')$$

$$= \frac{1}{N^2}\phi_w \sum \frac{1}{\pi_i} + V_{CP}(\overline{y}'')$$

$$= E_R V_{CP}(\overline{r}'') + V_R E_p(\overline{r}'')$$

$$= \frac{1}{N^2}E_R\left[\sum_{i<}^N\sum_j^N(\pi_i\pi_j - \pi_{ij})\left(\frac{r_i}{\pi_i} - \frac{r_j}{\pi_j}\right)^2 + V_R\left(\sum_1^N r_i\right)\right] + \sum \frac{r_i^2}{\pi_i}\beta_i$$

$$= \frac{1}{N^2}E_R E_p\left[\sum_{i<}\sum_j(\pi_i\pi_j - \pi_{ij})\left(\frac{r_i}{\pi_i} - \frac{r_j}{\pi_j}\right)^2\frac{I_{sij}}{\pi_{ij}} + \sum \frac{r_i^2}{\pi_i}\beta_i\frac{I_{si}}{\pi_i}\right] + \frac{\phi_w}{N}.$$

So,

$$v_3(\overline{r}'') = \frac{1}{N^2}\left[\sum_{i<}\sum_j(\pi_i\pi_j - \pi_{ij})\left(\frac{r_i}{\pi_i} - \frac{r_j}{\pi_j}\right)^2\frac{I_{sij}}{\pi_{ij}} + \sum \frac{r_i^2}{\pi_i}\beta_i\frac{I_{si}}{\pi_i}\right] + \frac{\phi_w}{N}$$

$$= v_{CP}(\overline{r}'') + \frac{\phi_w}{N}$$

and also

$$v_4(r'') = v_{CP}(r'') + \frac{\phi_w}{N^2}\sum \frac{I_{si}}{\pi_i}$$

are two more unbiased estimators of $V_{CP}(\bar{r}'')$. In case $\gamma(s)$ is a constant for every s with $p(s) > 0$, $v_5(\bar{r}'') = v_{YG}(r'') + \phi_w/N$ and $v_6(\bar{r}'') = v_{YG}(r'') + (\phi_w/N^2)$ $\sum I_{si}/\pi_i$, then there are two more unbiased estimators.

It is interesting to observe that in case of DR surveys $\bar{y}' = \sum_{i \in s_m} y_i/m$ is better than $\bar{y} = (1/n)\sum_{K=1}^{n} y_k$ as proved by D. Basu (1958) through a "sufficiency" argument to show that $V_p(\bar{y}') \le V_p(\bar{y})$ because the set of distinct units in an SRSWR with the respective variate values is the "minimal sufficient" statistic. Through algebraic arguments the same result has been demonstrated by Des Raj and Khamis (1958), Chikkagoudar (1966), Thionet (1967), Korwar and Serfling (1970), Lanke (1975a), Asok (1980) among others. Pathak (1962) also confirmed this and in addition he studied the relative efficiencies $\bar{y}$ versus $\bar{y}''$ and $\bar{y}'$ versus $\bar{y}''$. Chaudhuri and Pal (2008) extended the investigation to $\bar{r}$ versus $\bar{r}'$, $\bar{r}$ versus $\bar{r}''$ and $\bar{r}'$ versus $\bar{r}''$. Interestingly, the situation in DR does not carry forward to that in RR. So, they presented only a simulation-based numerical study.

Chaudhuri, Bose, and Dihidar's Estimators

Mangat et al. (1995) showed that

$$V(\bar{r}') < V(\bar{r})$$

if $N, n, p,$ and θ are such that

$$\theta(1 - \theta) > \frac{n(N - 1)(6N + n - 1)}{N\{6Nn - 12N - n(n - 1)\}} \frac{p(1 - p)}{(2p - 1)^2}.$$

It is hard to check or visualize circumstances favorable to this. Mangat et al. (1995) illustrated numerically that when $N = 100$, $n = 10$, $p = 0.9$, $V(\bar{r}') < V(\bar{r})$ if $0.236 \le \theta \le 0.764$ and $V(\bar{r}') < V(\bar{r})$ for other values of θ.

Chaudhuri and Pal (2008) illustrate two situations, namely

1. $N = 58, n = 13, \theta = 0.14, p = 0.45$
2. $N = 58, n = 5, \theta = 0.05, p = 0.45.$

   Under 1: they find $V(\bar{r}') < V(\bar{r}) < V(\bar{r}'')$
   Under 2: $V(\bar{r}') < V(\bar{r}) < V(\bar{r}'')$

Mangat et al. (1995) and Chaudhuri and Pal (2008) are common in the sense of employing Warner's (1965) RRT and in taking an SRSWR in a prespecified number of draws and they identify distinct persons in the chosen sample and gather just one RR from each. They all consider $\bar{r}$ and $\bar{r}'$. But Chaudhuri and Pal (2008) in addition bring $\bar{r}''$ into contention and they also present unbiased estimators for the variances, while Mangat et al. (1995) avoid taking that trouble.

One important point to note here is that in case of $\bar{r}$ additional RRs from each distinctly sampled person, as many times as the person appears in the sample are availed of, while in $\bar{r}'$ as well as in $\bar{r}''$ only a single RR is utilized with respect to each person sampled. Chaudhuri et al. (2009a) propose two additional estimators presented below utilizing repeated RRs gathered by Warner's device from each distinct person in the SRSWR independently and separately as many times as the person appears in the sample. Arnab (1999) also earlier advocated the use of these entire bodies of these RRs. His estimator is shown to outperform $\bar{r}'$ but could not be shown to beat $\bar{r}$.

Next we consider the treatment given by Chaudhuri et al. (2009a). Let s denote the sample chosen by SRSWR in n draws. Let f_i be the number of times the unit i appears in the sample s, $I_{ij} = 1/0$ accordingly as the ith person on his jth appearance in the sample gets a "match/mismatch" with the Warner's trial performed.

Then, $\Sigma_1^N f_i = n_i$; as before let s_m be the set of distinct units in s and m be their number $2 \le m \le n$. For $f_i > 0$, let

$$m_i = \frac{1}{f_i} \sum_{j=1}^{f_i} I_{ij}, \quad g_i = \frac{m_i - (1-p)}{(2p-1)}.$$

Then, $E_R(I_{ij}) = py_i + (1-p)(1-y_i) = (1-p) + (2p-1)y_i = E_R(m_i),$

$$V_R(I_{ij}) = E_R(I_{ij})(1 - E_R(I_{ij})) = p(1-p),$$

$$V_R(m_i) = \frac{p(1-p)}{f_i}, \quad E_R(g_i) = y_i, \quad V_R(g_i) = \frac{\phi_w}{f_i}.$$

Chaudhuri et al. (2009a) propose two estimators for θ based on g_i for $i \in s_m$, namely $t_1 = (1/m)\Sigma_{i \in s_m} g_i = \bar{r}(m)$, and

$$t_2 = \frac{1}{N} \sum_{i \in s_m} \frac{g_i}{\pi_i}.$$

Then, $E(t_1) = E_p E_R(t_1)$

$$= E_p(\bar{y}') = E_p\left(\frac{1}{m} \sum_{j \in s_m} y_j\right) = \bar{Y} = \theta.$$

As noted in this chapter under "Chaudhuri and Pal's Estimators,"

$$S^2 = \frac{1}{N-1} \sum_1^N (y_i - \bar{Y})^2 = \frac{N\theta(1-\theta)}{N-1}$$

Pathak (1962) showed that

$$V_p\left(\bar{y}'\right) = \left[E_p\left(\frac{1}{\gamma}\right) - \frac{1}{N}\right]S^2, \quad E_p\left(\frac{1}{\gamma}\right) = \frac{1^{n-1} + \cdots + N^{n-1}}{N},$$

and

$$V_p\left(m\bar{y}'\right) = N\theta(1-\theta)\left[\left(1 - \frac{1}{N}\right)^n - \left(1 - \frac{2}{N}\right)^n\right]$$

$$+ \theta^2 N\left[\left(1 - \frac{1}{N}\right)^n - N\left(1 - \frac{1}{N}\right)^{2n} + (N-1)\left(1 - \frac{2}{N}\right)^n\right]$$

So,

$$V(t_1) = V_p\left(\bar{y}(m)\right) + E_p\left(\frac{\phi_w}{m^2}\sum_{i \in s_m}\frac{1}{f_i}\right)$$

$$= \left[E_p\left(\frac{1}{m}\right) - \frac{1}{N}\right]S^2 + \phi_w E_p\left(\frac{1}{m^2}\sum_{i \in s_m}\frac{1}{f_i}\right)$$

$$= \left[\frac{1}{N^{n-1}(N-1)}\sum_{j=1}^{N-1}j^{n-1}\right]\theta(1-\theta) + \phi_w E_p\left(\frac{1}{m^2}\sum_{i \in s_m}\frac{1}{f_i}\right),$$

$$E(t_2) = E_p E_R\left(t_2\right) = E_p\left(\frac{1}{N}\sum_{i \in s_m}\frac{y_i}{\pi_i}\right) = \bar{Y} = \theta,$$

$$V(t_2) = \frac{\phi_w}{N^2\pi_i^2}E_p\left(\sum_{i \in s_m}\frac{1}{f_i}\right) + \frac{1}{N^2\pi_i^2}V_p\left(m\bar{y}'\right),$$

because $\pi_i = 1 - (1 - 1/N)^n$ for every $i = 1, \ldots, N$.
  So,

$$V(t_2) = \frac{\phi_w}{N^2\pi_i^2}E_p\left(\sum_{i \in s_m}\frac{1}{f_i}\right) + \frac{\theta(1-\theta)}{N\pi_i^2}\left[\left(1 - \frac{1}{N}\right)^n - \left(1 - \frac{2}{N}\right)^n\right]$$

$$+ \frac{\theta^2}{N^2\pi_i^2}\left[N\left(1 - \frac{1}{N}\right)^n - N^2\left(1 - \frac{1}{N}\right)^{2n} + N(N-1)\left(1 - \frac{2}{N}\right)^n\right].$$

A comparison of the relative performance of $\bar{r}$ versus $\bar{r}'$ versus t_1 versus t_2 may be quoted here from the works of Chaudhuri et al. (2009a).

For $1 \le f_i \le n \,\forall i$ it follows that $\sum_{i \in s_m} 1/f_i \le m$ but $\sum_{i \in s_m} 1/f_i \ge m/n$. Clearly, $V(t_1) \le V(\bar{r}')$, that is, $t_1 > \bar{r}'$.

It is gratifying that algebra supports this because intuition demands it. In $\bar{r}'$ only one RR per distinct person sampled is utilized while t_1 utilizes an independently realized RR from each distinct person as many times as the person appears in the sample and hence uses additional data.

On the other hand,

$$V\left(\bar{r}'\right) - V\left(t_2\right) = \phi_w\left[E_p\left(\frac{1}{m}\right) - \frac{1}{N^2\pi_i^2}E_p\left(\sum_{i\in s_m}\frac{1}{f_i}\right)\right] + \frac{N\theta(1-\theta)}{(N-1)}A_1 - \theta^2 A_2,$$

when

$$A_1 = \frac{1}{N^n}\sum_{j=1}^{N-1}j^{n-1} - \frac{N-1}{N^2\pi_i^2}\left[\left(1-\frac{1}{N}\right)^n - \left(1-\frac{2}{N}\right)^n\right],$$

$$A_2 = \frac{1}{N^2\pi_i}\left[\begin{array}{c}\left(1-\dfrac{1}{N}\right)^n - N\left(1-\dfrac{1}{N}\right)^{2n} \\[2mm] + (N-1)\left(1-\dfrac{2}{N}\right)^n\end{array}\right].$$

Korwar and Serfling (1970) considered

$$Q = \frac{1}{n} + \frac{1}{2N} + \frac{n-1}{12N^2},$$

and showed that for $n \geq 3$,

$$Q - \frac{1}{720N} < E_p\left(\frac{1}{m}\right) \leq Q.$$

Then follows

$$V\left(\bar{r}'\right) - V\left(t_2\right) \leq \phi_w\left[Q - \frac{E_p(m)}{nN^2\pi_i^2}\right] + \frac{N\theta(1-\theta)}{(N-1)}A_1 - \theta^2 A_2,$$

that is,

$$V\left(\bar{r}'\right) - V\left(t_2\right) \leq \phi_w\left[Q - \frac{1}{nN\pi_i}\right] + \frac{N\theta(1-\theta)}{(N-1)}A_1 - \theta^2 A_2.$$

So,

$$V(\bar{r}') \leq V(t_2) \quad \text{if} \quad \frac{N\theta(1-\theta)}{(N-1)}A_1 - \theta^2 A_2 \leq \phi_w\left(\frac{1}{nN\pi_i} - Q\right).$$

Also,

$$V\left(\bar{r}'\right) - V\left(t_2\right) \geq \phi_w\left[Q - \frac{1}{720N} - \frac{E_p(m)}{N^2\pi_i^2}\right] + \frac{N\theta(1-\theta)}{(N-1)}A_1 - \theta^2 A_2.$$

Thus,

$$V(\bar{r}') - V(t_2) > \phi_w \left[Q - \frac{1}{720N} - \frac{1}{N\pi_i} \right] + \frac{N\theta(1-\theta)}{(N-1)} A_1 - \theta^2 A_2.$$

So,

$$V(t_2) < V(\bar{r}') \quad \text{if} \quad \frac{N\theta(1-\theta)}{(N-1)} A_1 - \theta^2 A_2 \geq \phi_w \left[\frac{1}{N\pi_i} - Q + \frac{1}{720N} \right].$$

Naturally if we neglect $1/720\,N$, then

$$V(t_2) < V(\bar{r}') \quad \text{if} \quad \frac{N\theta(1-\theta)}{(N-1)} A_1 - \theta^2 A_2 \geq \phi_w \left[\frac{1}{N\pi_i} - Q \right].$$

So, $t_2 > \bar{r}'$ under one condition and $t_2 < \bar{r}'$ under another condition as above. Chaudhuri et al. (2009a) illustrate more situations to facilitate comparison of t_2 versus $\bar{r}'$.

Next they observe that

$$V(t_1) - V(\bar{r}) = \phi_w \left[E_p \left(\frac{1}{m^2} \sum \frac{1}{f_i} \right) - \frac{1}{n} \right] + \theta(1-\theta) \left[\frac{NE_p(1/m) - 1}{N-1} - \frac{1}{n} \right]$$

$$\leq \phi_w \left[Q - \frac{1}{n} \right] + \theta(1-\theta) \left[\frac{NQ-1}{N-1} - \frac{1}{n} \right].$$

So,

$$V(t_1) \leq V(\bar{r}) \quad \text{if} \quad \theta(1-\theta) \geq \phi_w (N-1) \left(\frac{nQ-1}{N+n-1-NnQ} \right).$$

They further show that

$$V(t_1) - V(\bar{r}) > \frac{\phi_w}{n} \left[Q - \frac{1}{720N} - 1 \right] + \theta(1-\theta) \left[\frac{NQ-(721/720)}{N-1} - \frac{1}{n} \right].$$

This implies that $V(t_1) > V(\bar{r})$ if

$$\theta(1-\theta) \leq \phi_w (N-1) \left(\frac{Q-(1/720N)-1}{N+(721n/720)-1-NnQ} \right).$$

It is very hard to find or visualize values of θ satisfying this inequality so that only on rare situations $\bar{r}$ may beat t_1.

Next, we see

$$V(t_2) - V(\bar{r}) = \phi_w \left[\frac{1}{N^2 \pi_i^2} E_p \left(\sum_{i \in s_m} \frac{1}{f_i} \right) - \frac{1}{n} \right] + V_p \left[\frac{m}{N\pi_i} \bar{y}' \right] - V_p(\bar{y})$$

$$\leq \phi_w \left(\frac{1}{N\pi_i} - \frac{1}{n} \right) + V_p \left(\frac{m\bar{y}'}{N\pi_i} \right) - V_p(\bar{y}),$$

since $E_p(\sum_{i \in s(m)} 1/f_i) \leq E_p(m) = N\pi_i$ with $i = 1, \ldots, N$.

So,

$$V(t_2) \leq V(\bar{r}) \quad \text{if} \quad V_p \left(\frac{m\bar{y}'}{N\pi_i} \right) - V_p(\bar{y}) + \phi_w \left(\frac{n - N\pi_i}{nN\pi_i} \right) \leq 0.$$

Using algebra it is possible to note $V(t_2) \leq V(\bar{r})$ if

$$\theta^2 \left[A_2 + \frac{NA_1}{N-1} - \frac{1}{N-1} \sum_{j=1}^{N-1} \left(\frac{j}{N} \right)^{n-1} + \frac{1}{n} \right]$$

$$+ \theta \left[\frac{1}{N-1} \sum_{j=1}^{N-1} \left(\frac{j}{N} \right)^{n-1} \frac{NA_1}{N-1} - \frac{1}{n} \right] + \phi_w \left(\frac{n - N\pi_i}{nN\pi_i} \right) \leq 0.$$

Similarly, on noting

$$E_p \left(\sum_{i \in s(m)} \frac{1}{f_i} \right) \geq \frac{E_p(m)}{n} = \frac{N\pi_i}{n},$$

it follows that $V(\bar{r}) \leq V(t_2)$ if

$$\theta^2 \left[A_2 + \frac{NA_1}{N-1} - \frac{1}{N-1} \sum_{j=1}^{N-1} \left(\frac{j}{N} \right)^{n-1} + \frac{1}{n} \right] + \theta \left[\frac{1}{N-1} \sum_{j=1}^{N-1} \left(\frac{j}{N} \right)^{n-1} - \frac{NA_1}{N-1} - \frac{1}{n} \right]$$

$$+ \frac{\phi_w}{n} \left(\frac{1 - N\pi_i}{N\pi_i} \right) \geq 0.$$

Finally,

$$V(t_1) - V(t_2) = \phi_w \left[E_p \left(\frac{1}{m^2} \sum_{i \in s(m)} \frac{1}{f_i} \right) - \frac{1}{N^2 \pi_i^2} E_p \left(\sum_{i \in s(m)} \frac{1}{f_i} \right) \right]$$

$$+ \frac{N\theta(1 - \theta)}{N-1} A_1 - \theta^2 A_2.$$

So, $V(t_1) \le V(t_2)$ whenever $V(\bar{r}') \le V(t_2)$.

Again neglecting $1/720Nn$, $V(t_1) > V(t_2)$ if

$$\frac{N\theta(1-\theta)A_1}{N-1} - \theta^2 A_2 \ge \phi_w\left(\frac{1}{N\pi_i} - \frac{Q}{n}\right).$$

For $\bar{r}, \bar{r}'$ unbiased variance estimators have been already presented. For $V(t_1)$ and $V(t_2)$ unbiased estimators are derived by Chaudhuri et al. (2009a) as follows.

Consider the notations

$$C_{1a} = (N-1)\left[\left(1-\frac{1}{N}\right)^a - \left(1-\frac{2}{N}\right)^n\right],$$

$$C_{1b} = N\left(1-\frac{1}{N}\right)^n - N^2\left(1-\frac{1}{N}\right)^{2n} + N(N-1)\left(1-\frac{2}{N}\right)^n.$$

Since r_i, r_j's $(i \ne j)$ are independent, using Pathak's (1962) results we get

$$E_p E_R\left(\sum_{i\ne}\sum_{j\in s(m)} r_i r_j\right) = E_p\left(\sum_{i\ne}\sum_{j\in s(m)} y_i y_j\right)$$

$$= \theta\left[C_{1a}\frac{N}{N-1} - N\pi_i\right] + \theta^2\left[C_{1b} + N^2\pi_i^2 - C_{1a}\frac{N}{N-1}\right].$$

So, an unbiased estimator for θ^2 is

$$\hat{\theta}^2 = \frac{1}{C_{1b} + N^2\pi_i^2 - C_{1a}\left(\frac{N}{N-1}\right)}\left[\sum_{i\ne}\sum_j r_i r_j - \bar{r}'\left(C_{1a}\frac{N}{N-1} - N\pi_i\right)\right]$$

So, an unbiased estimator for $V(\bar{r}')$ is

$$v_1(\bar{r}') = \phi_w E_p\left(\frac{1}{m}\right) + \left[\frac{NE_p(1/m)-1}{(N-1)}\right]\left(\bar{r}' - \hat{\theta}^2\right)$$

and the other is

$$v_2(\bar{r}') = \phi_w\left(\frac{1}{m}\right) + \left[\frac{NE_p(1/m)-1}{(N-1)}\right]\left(\bar{r}' - \hat{\theta}^2\right)$$

in addition to those earlier presented.

Proceeding similarly,

$$E_p E_R \left(\sum_{i \neq} \sum_j g_i g_j \right) = \theta \left[C_{1a} \frac{N}{N-1} - N\pi_i \right] + \theta^2 \left[C_{1b} + N^2 \pi_i^2 - C_{1a} \frac{N}{N-1} \right]$$

So, an unbiased estimator for θ^2 is

$$\tilde{\theta}^2 = \frac{1}{C_{1b} + N^2 \pi_i^2 - C_{1a} \left(\dfrac{N}{N-1} \right)} \left[\sum_{i \neq j} \sum_{\in s(m)} g_i g_j - t_1 \left(C_{1a} \frac{N}{N-1} - N\pi_i \right) \right]$$

So,

$$\left[\frac{1}{N^{n-1}(N-1)} \sum_{j=1}^{N-1} j^{n-1} \right] \left(t_1 - \tilde{\theta}^2 \right),$$

is an unbiased estimator of

$$\left[E_p \left(\frac{1}{m} \right) - \frac{1}{N} \right] \frac{N\theta(1-\theta)}{(N-1)}.$$

So, an unbiased estimator of $V(t_1)$ is

$$v_1(t_1) = \left[\frac{1}{N^{n-1}(N-1)} \sum_{j=1}^{N-1} j^{n-1} \right] \left(t_1 - \tilde{\theta}^2 \right) + \phi_w \left(\frac{1}{m^2} \right) \sum_{i \in s(m)} \frac{1}{f_i}$$

To derive a few more, consider the following:

$$v_r(m) = \frac{1}{(m-1)} \sum_{i \in s(m)} \left(g_i - \bar{r}' \right)^2, \quad v(m) = \frac{1}{(m-1)} \sum_{i \in s(m)} \left(y_i - \bar{y}' \right)^2,$$

$$C_m(n) = \sum_0^{m-1} (-1)^i \binom{m}{i} (m-i)^n,$$

$$C_2 = \frac{1}{N^n} \left(\sum_{j=1}^{N-1} j^{n-1} \right) \left(\frac{N}{N-1} \right) \frac{C_m(n) - C_m(n-1)}{C_m(n)},$$

$$v_2(\bar{r}') = c_2 v_r(m),$$

$$C_3 = \frac{C_{m-1}(n-1)}{C_m(n)}, \quad C_4 = \left[\left(\frac{1}{m} - \frac{1}{N} \right) + \left(\frac{N-1}{N^n - N} \right) \right],$$

$$C_5 = \left[\left(\frac{1}{m} - \frac{1}{N} \right) + N^{1-n} \left(1 - \frac{1}{m} \right) \right].$$

Then, using

$$E_p\left[C_i v(m)\right] = \left[E_p\left(\frac{1}{m}\right) - \frac{1}{N}\right]S^2,$$

$i = 2, \ldots, 5$ from Pathak (1962), on simplification one may observe

$$E_p E_R\left[v_2\left(\bar{r}'\right)\right] = c_2 E_p E_R\left[\left(\frac{1}{m-1}\right)\sum_{i\in s(m)}\left(g_i - \frac{\sum_{i\in s(m)}g_i}{m}\right)^2\right]$$

$$= c_2 E_p E_R\left[\frac{1}{m-1}\left(\sum_{i\in s(m)}g_i^2 - m\left(\bar{g}\right)^2\right)\right],$$

$$\bar{g} = \frac{\sum_{i\in s(m)}g_i}{m} = C_2 E_p\left[\left(\frac{1}{m-1}\right)\sum_{i\in s(m)}\left\{V_R\left(g_i\right) + \left(E_R\left(g_i\right)\right)^2\right\}\right.$$

$$\left. -\left(\frac{m}{m-1}\right)\left\{V_R\left(\bar{g}\right) + \left(E_R\left(\bar{g}\right)\right)^2\right\}\right]$$

$$= C_2 E_p\left[\frac{1}{(m-1)}\sum_{i\in s(m)}\left(\frac{\phi_w}{f_i} + y_i^2\right) - \left(\frac{m}{m-1}\right)\left(\frac{\phi_w}{m^2}\sum_{i\in s(m)}\frac{1}{f_i} + \bar{y}^2\right)\right]$$

$$= C_2 E_p\left[\frac{\phi_w}{m}\sum_{i\in s(m)}\frac{1}{f_i} + \frac{1}{(m-1)}\left\{\sum_{i\in s(m)}y_i^2 - m\bar{y}^2\right\}\right]$$

$$= C_2 E_p\left[\frac{\phi_w}{m}\sum_{i\in s(m)}\frac{1}{f_i}\right] + E_p\left[C_2 v(m)\right]$$

$$= E_p E_R\left[C_2\frac{\phi_w}{m}\sum_{i\in s(m)}\frac{1}{f_i}\right] + \left[E_p\left(\frac{1}{m}\right) - \frac{1}{N}\right]S^2.$$

So,

$$\left[v_2\left(\bar{r}'\right) - C_2\frac{\phi_w}{m}\sum_{i\in s(m)}\frac{1}{f_i}\right]$$

is an unbiased estimator of

$$V(\bar{y}') = \left[E_p\left(\frac{1}{m}\right) - \frac{1}{N}\right] S^2.$$

Also, another unbiased estimator of $V(t_1)$ is

$$v(t_1) = \frac{\phi_w}{m^2} \sum_{i \in s(m)} \frac{1}{f_i} + v_2(\bar{r}') - C_2 \frac{\phi_w}{m} \sum_{i \in s(m)} \frac{1}{f_i}$$

$$= \frac{\phi_w}{m}\left(\frac{1}{m} - C_2\right) \sum_{i \in s(m)} \frac{1}{f_i} + v_2(\bar{r}') = v_2(t_1).$$

Similarly, three more unbiased estimators of $V(t_1)$ are

$$v_i(t_1) = \frac{\phi_w}{m}\left(\frac{1}{m} - C_i\right) \sum_{i \in s(m)} \frac{1}{f_i} + C_i v_i(m), \quad i = 3, 4, 5.$$

In order to derive unbiased estimators from Chaudhuri et al. (2009a) we quote as follows:
   Let

$$v_{HT}(g) = \frac{1}{N^2} \left[\sum_{i \in s_m} g_i^2 \left(\frac{1 - \pi_i}{\pi_i^2}\right) + \sum_{i \neq} \sum_{i'} g_i g_{i'} \left(\frac{\pi_{ii'} - \pi_i \pi_{i'}}{\pi_{ii'} \pi_i \pi_{i'}}\right)\right].$$

Then,

$$Ev_{HT}(g) = \frac{1}{N^2} \left[\sum y_i^2 \left(\frac{1 - \pi_i}{\pi_i}\right) + \phi_w E_p \left(\sum_{i \in s(m)} \frac{1}{f_i}\left(\frac{1 - \pi_i}{\pi_i^2}\right)\right)\right.$$

$$\left. + \sum_{i \neq} \sum_{i'} y_i y_{i'} \left(\frac{\pi_{ii'} - \pi_i \pi_{i'}}{\pi_i \pi_{i'}}\right)\right]$$

$$= V_p \left(\frac{1}{N} \sum_{i \in s(m)} \frac{y_i}{\pi_i}\right) + \frac{\phi_w}{N^2} E_p \left(\sum_{i \in s(m)} \frac{1}{f_i}\left(\frac{1 - \pi_i}{\pi_i^2}\right)\right)$$

$$= V_p \left[E_R(t_2)\right] + \frac{\phi_w}{N^2} E_p \left(\sum_{i \in s(m)} \frac{1}{f_i}\left(\frac{1 - \pi_i}{\pi_i^2}\right)\right)$$

$$= v(t_2) - E_p \left[V_R(t_2)\right] + \frac{\phi_w}{N^2} E_p \left(\sum_{i \in s(m)} \frac{1}{f_i}\left(\frac{1 - \pi_i}{\pi_i^2}\right)\right).$$

So,

$$V(t_2) = E_p E_R \left(v_{HT}(g) \right) + E_p \left[\frac{\phi_w}{N^2} \sum_{i \in s(m)} \frac{1}{\pi_i^2 f_i} \right] - \frac{\phi_w}{N^2} E_p \left[\sum_{i \in s(m)} \frac{1}{f_i} \left(\frac{1 - \pi_i}{\pi_i^2} \right) \right].$$

So,

$$v(t_2) = v_{HT}(g) + \frac{\phi_w}{N^2} \sum_{i \in s(m)} \frac{1}{\pi_i f_i}$$

is an unbiased estimator of $V(t_2)$.

More are readily available on considering the forms $V_{CP}(\bar{r}'')$.

Inverse SRSWR

Chaudhuri, Bose, and Dihidar's Estimators

SRSWR continued till the first time a preassigned number of distinct individuals appear in the sample drawn, is known as Inverse SRSWR (Chaudhuri et al., 2009b).

In this section, we present unbiased estimators for θ along with unbiased variance estimators. Especially, each time a person is selected, the person, on request, produces an RR by Warner's RR device in independent manners.

Suppose $\gamma(>1)$ is the preassigned number of distinct units to be sampled in an inverse SRSWR scheme. Suppose the 1st, 2nd, ..., $(\gamma - 1)$th distinct person appears $f_{1s}, f_{2s}, ..., f_{(\gamma-1)s}$ times, respectively with these integers being random variables. Clearly, the γth distinct person appears only once. If n is the number of random draws needed to yield these γ distinct persons, then

$$\sum_{i=1}^{\gamma-1} f_{is} = (n - 1). \tag{2.1}$$

Des Raj and Khamis (1958) have given the probability distribution of this random number n as

$$P(n) = \frac{N \binom{N-1}{\gamma-1}}{N^n} \sideset{}{'}\sum \frac{(n-1)!}{f_{1s}! \cdots f_{(\gamma-1)s}!}$$

$$= \frac{\binom{N-1}{\gamma-1}}{N^{n-1}} \left[\Delta^{\gamma-1} x^{n-1} \big|_{x=0} \right] \tag{2.2}$$

Here, Σ' denotes the sum over all possible positive integers $f_{is}(i = 1, \ldots, (\gamma - 1))$ subject to (Equation 2.1) and Δ is the difference operator such that $\Delta f(x) = f(x + 1) - f(x)$.

As a consequence,

$$\sum{}' \frac{(n-1)!}{f_{1s}! \ldots f_{(\gamma-1)s}!} = (\gamma - 1)^{n-1} - (\gamma - 1)(\gamma - 2)^{n-1} + \cdots + (-1)^{\gamma-2}\binom{\gamma - 1}{\gamma - 2}$$

$$= [\Delta^{\gamma-1}x^{n-1}]_{x=0}.$$

We may write $E_p = E_n E_{p|n}$, $V_p = E_n V_{p|n} + V_n E_{p|n}$ with $E_{p|n}$, $V_{p|n}$ as expectation-variance operators conditional on a specified n and E_n, V_n as the expectation-variance operators over the distribution of n.

Based on the inverse SRSWR requiring n draws to produce γ distinct persons, let an estimator for θ analogous to Warner's $\bar{r}$ based on SRSWR with fixed n as the number of draws be

$$e_1 = \left(\frac{1}{n}\right)\sum_{k=1}^{n} r_k = \bar{r}. \tag{2.3}$$

Then,

$$E(e_1) = E_n E_{p|n}\left[E_R(e_1)\right] = \frac{1}{N}\sum_{1}^{N} y_i = \theta.$$

$$V(e_1) = E_n\left[V_{p|n}E_R(e_1) + E_{p|n}V_R(e_1)\right]$$

$$= E_n\left[V_{p|n}(\bar{y}_n)\right] + \phi_w E_n\left(\frac{1}{n}\right).$$

Chikkagoudar (1966) has given

$$E_n\left(\frac{1}{n}\right) = \sum_{n=\gamma}^{\alpha} \frac{1}{n}P(n) = \left(\frac{N-1}{\gamma-1}\right)\sum_{n=\gamma}^{\infty} \frac{N^{1-n}}{n}\left[\Delta^{\gamma-1}x^{n-1}\big|_{x=0}\right],$$

$$V_{p|n}(\bar{y}_n) = \left[\left(\frac{N-n}{Nn}\right) + \frac{(n-1)(n-2)}{n^2}\frac{\left[\Delta^{\gamma-1}x^{n-2}\big|_{x=0}\right]}{\left[\Delta^{\gamma-1}x^{n-1}\big|_{x=0}\right]}\right]S^2,$$

$$E_n V_{p|n}(\bar{y}_n) = S^2\left(\frac{N-1}{\gamma-1}\right)\left[\Delta^{\gamma-1}\left\{\frac{1}{x} + \frac{N(x-3)}{x^2}\log\left(\frac{N}{N-x}\right) + \frac{2}{x}\sum_{n=\gamma}^{\alpha}\frac{1}{n^2}\left(\frac{x}{N}\right)^{n-1}\right\}\big|_{x=0}\right].$$

Writing $(1/N)\sum_{1}^{N}(y_i - \bar{Y})^2 = \theta(1 - \theta)$, Lanke (1975a) showed

$$E_n V_{p|n}(\bar{y}_n) = \frac{\sigma^2}{(N-1)}\left[NE_n\left(\frac{1}{n}\right) - E_n\left(\frac{3n+1}{(n+1)^2}\right)\right].$$

Using these results two variance formulae follow as

$$V(e_1) = \frac{N}{N-1}\theta(1-\theta)\binom{N-1}{\gamma-1}$$

$$\times\left[\Delta^{\gamma-1}\left\{\frac{1}{x}+\frac{N(x-3)}{x^2}\log\left(\frac{N}{N-x}\right)+\frac{2}{x}\sum_{n=\gamma}^{\alpha}\frac{1}{n^2}\left(\frac{x}{N}\right)^{n-1}\right\}\bigg|_{x=0}\right]+\phi_w E_n\left(\frac{1}{n}\right)$$

and

$$V'(e_1) = \frac{\theta(1-\theta)}{(N-1)}\left[NE_n\left(\frac{1}{n}\right)-E_n\left(\frac{3n+1}{(n+1)^2}\right)\right]+\phi_w E_n\left(\frac{1}{n}\right).$$

Suppose in the inverse SRSWR from the population of N units requiring n draws yielding γ distinct units as in Mangat et al. (1995) plan with each distinctly drawn person in the set u of γ distinct persons executing Warner's RR technique only once each one may employ the estimator for θ: $e_2 = (1/\gamma)\sum_{i\in u} r_i$.

As against this, one may employ the alternative: $e_3 = (1/\gamma)\sum_{i\in u}\hat{y}_i$.

Here $\hat{y}_i$ denotes the following:

$I_{ij} = 1/0$ as in the jth RRT the ith distinct person gets a "match"/ "mis-match";

$j = 1, \ldots, f_{is}$;

$$u_i = \frac{1}{f_{is}}\sum_{j=1}^{f_{is}} I_{ij}, \quad \hat{y}_i = \frac{u_i - (1-p)}{(2p-1)}.$$

It easily follows that

$$E_R(\hat{y}_i) = y_i, \quad E(e_2) = \theta = E(e_3), \quad V_R(\hat{y}_i) = \frac{\phi_w}{f_{is}}.$$

Next, follow

$$V(e_2) = V_p\left(\frac{1}{\gamma}\sum_{i\in u} y_i\right)+E_p\left(\frac{\phi_w}{\gamma}\right)$$

$$= \left(\frac{1}{\gamma}-\frac{1}{N}\right)\frac{N}{N-1}\theta(1-\theta)+\frac{\phi_w}{\gamma},$$

$$V(e_3) = V_p\left(\frac{1}{\gamma}\sum_{i\in u} y_i\right)+E_p\left(\frac{1}{\gamma^2}\phi_w\sum_{i\in u}\frac{1}{f_{is}}\right)$$

$$= \left(\frac{1}{\gamma}-\frac{1}{N}\right)\frac{N}{N-1}\theta(1-\theta)+\frac{\phi_w}{\gamma^2}E_n E_{p|n}\left(\sum_{i\in u}\frac{1}{f_{is}}\right).$$

Finally for this inverse SRSWR let us consider the Horvitz–Thompson estimators

$$e_4 = \frac{1}{N} \sum_{i \in u} \frac{r_i}{\pi_i} = \frac{\gamma}{N \pi_i} e_2$$

and

$$e_5 = \frac{1}{N} \sum_{i \in u} \frac{\hat{y}_i}{\pi_i} = \frac{\gamma}{N} e_3,$$

respectively, based on r_i's and $\hat{y}_i$'s as above. Here π_i, the inclusion probability in this scheme is

$$\pi_i = 1 - \left[\sum_{n=\gamma}^{\alpha} P(n) \frac{(N-1)\binom{N-2}{\gamma-1}}{N\binom{N-1}{\gamma-1}} = \frac{N-\gamma}{N} \right],$$

since

$$\sum_{n=\gamma}^{\alpha} P(n) = 1.$$

So,

$$\pi_i = \frac{\gamma}{N} \Rightarrow e_4 = e_2 \quad \text{and} \quad e_3 = e_5.$$

Since

$$\sum_{i \in u} \frac{1}{f_{is}} \le \gamma; \quad \sum_{i \in u} \frac{1}{f_{is}} \ge \frac{\gamma}{n}, \tag{2.4}$$

it follows that $V(r_3) \le V(e_2)$, that is $e_3 > e_2$ uniformly. To compare e_1 versus e_3, we note that

$$
\begin{aligned}
V(e_3) - V'(e_1) &= \left(\frac{1}{\gamma} - \frac{1}{N} \right) \frac{N\theta(1-\theta)}{(N-1)} + \frac{\phi_w}{\gamma^2} E_n E_{p|n} \left(\sum_{i \in u} \frac{1}{f_{is}} \right) \\
&\quad - \left[\frac{\theta(1-\theta)}{(N-1)} \left\{ NE_n \left(\frac{1}{n} \right) - E_n \left(\frac{3n+1}{(n+1)^2} \right) \right\} + \phi_w E_n \left(\frac{1}{n} \right) \right] \\
&< \left(\frac{1}{\gamma} - \frac{1}{N} \right) \frac{N}{N-1} \theta(1-\theta) + \frac{\phi_w}{\gamma} - \frac{\theta(1-\theta)}{(N-1)} NE_n \left(\frac{1}{n} \right) \\
&\quad + \frac{\theta(1-\theta)}{N-1} \frac{3}{\gamma} - \phi_w E_n \left(\frac{1}{n} \right)
\end{aligned}
$$

by Equation 2.4 and since

$$E_n\left(\frac{3n+1}{(n+1)^2}\right) < \frac{3}{\gamma}.$$

Moreover, Lanke (1975a) has shown that if $\gamma \to \infty$, $N \to \infty$, $\gamma/N \to f_0 (0 < f_0 < 1)$, then

$$NE_n\left(\frac{1}{n}\right) \to \frac{1}{\log(1/1 - f_0)}.$$

So, for large N,

$$V(e_3) - V'(e_1) < \left(\frac{1}{\gamma} - \frac{1}{N}\right)\frac{N}{N-1}\theta(1-\theta) + \frac{\phi_w}{\gamma} + \frac{\theta(1-\theta)}{N-1}\frac{3}{\gamma}\frac{\left[\frac{\theta(1-\theta)}{N-1} + \frac{\phi_w}{N}\right]}{\log\left(\frac{1}{1-f_0}\right)}.$$

Applying algebra it follows that $V(e_3) < V'(e_1)$ if

$$\theta(1-\theta) \geq \frac{N-1}{N}\phi_w\left[\frac{N\log\left(\frac{1}{1-f_0}\right) - \gamma}{\gamma - (N+3-\gamma)\log\left(\frac{1}{1-f_0}\right)}\right].$$

To derive unbiased variance estimators, we write

$$V(e_1) = E_n\left[\left\{\left(\frac{N-n}{Nn}\right) + \frac{(n-1)(n-2)}{n^2}\frac{\left[\Delta^{\gamma-1}x^{n-2}\big|_{x=0}\right]}{\left[\Delta^{\gamma-1}x^{n-1}\big|_{x=0}\right]}\right\} + \frac{\phi_w}{n}\right].$$

Repeatedly using Chikkagoudar's (1966) results and recalling

$$\bar{r}_n = \frac{1}{n}\sum_{k=1}^{n}r_k = e_1, \quad s_n^2 = \frac{1}{(n-1)}\sum_{k=1}^{n}(y_k - \bar{y}_n)^2,$$

we get

$$E_R\left[\frac{1}{n(n-1)}\sum_{k=1}^{n}(r_k - \bar{r}_n)^2\right]$$

$$= \frac{1}{n(n-1)} \left[\sum_{k=1}^{n} \{V_R(r_k) + E_R^2(r_k)\} - n\{V_R(\bar{r}_n) + E_R^2(\bar{r}_n)\} \right]$$

$$= \frac{1}{n(n-1)} \left[n\phi_w + \sum_{k=1}^{n} y_k^2 - n\frac{\phi_w}{n} - n\bar{y}_n^2 \right] = \frac{\phi_w}{n} + \frac{s_n^2}{n}.$$

Then,

$$E_{p|n} E_R \left[\frac{1}{n(n-1)} \sum_{k=1}^{n} (r_k - \bar{r}_n)^2 \right]$$

$$= \frac{\phi_w}{n} + \frac{1}{n} E_{p|n}(s_n^2)$$

$$= \frac{\phi_w}{n} + \frac{1}{n} \left[1 - \frac{n-2}{n} \frac{\left[\Delta^{\gamma-1} x^{n-2} \big|_{x=0} \right]}{\left[\Delta^{\gamma-1} x^{n-1} \big|_{x=0} \right]} \right] S^2.$$

Since Chikkagouder (1966) showed that

$$E_{p|n}(s_n^2) = \left[1 - \frac{n-2}{n} \frac{\left[\Delta^{\gamma-1} x^{n-2} \big|_{x=0} \right]}{\left[\Delta^{\gamma-1} x^{n-1} \big|_{x=0} \right]} \right] S^2$$

and so, S^2 is unbiasedly estimated by

$$\left[1 - \frac{n-2}{n} \frac{\left[\Delta^{\gamma-1} x^{n-2} \big|_{x=0} \right]}{\left[\Delta^{\gamma-1} x^{n-1} \big|_{x=0} \right]} \right]^{-1} \left[\frac{1}{(n-1)} \sum_{k=1}^{n} (r_k - \bar{r}_n)^2 - \phi_w \right].$$

Hence, an unbiased estimator for $V(e_1)$ is

$$\hat{V}(e_1) = \frac{\phi_w}{n} + \left\{ \left(\frac{N-n}{Nn} \right) + \frac{(n-1)(n-2)}{n^2} \frac{\left[\Delta^{\gamma-1} x^{n-2} \big|_{x=0} \right]}{\left[\Delta^{\gamma-1} x^{n-1} \big|_{x=0} \right]} \right\}$$

$$\times \left[1 - \frac{n-2}{n} \frac{\left[\Delta^{\gamma-1} x^{n-2} \big|_{x=0} \right]}{\left[\Delta^{\gamma-1} x^{n-1} \big|_{x=0} \right]} \right]^{-1} \times \left[\frac{1}{(n-1)} \sum_{k=1}^{n} (r_k - \bar{r}_n)^2 - \phi_w \right].$$

To find an unbiased estimator for $V(e_2)$, note

$$E_R \left[\left(\frac{1}{\gamma} - \frac{1}{N} \right) \frac{1}{(\gamma - 1)} \sum_{i \in u} (r_i - e_2)^2 \right]$$

$$= \left(\frac{1}{\gamma} - \frac{1}{N} \right) \frac{1}{(\gamma - 1)} \left[\sum_{i \in u} E_R(r_i^2) - \gamma E_R(e_2)^2 \right]$$

$$= \left(\frac{1}{\gamma} - \frac{1}{N}\right)\frac{1}{(\gamma - 1)}\left[\sum_{i \in u}\left\{V_R(r_i) + \left(E_R(r_i)\right)^2\right\} - \gamma\left\{V_R(e_2) + \left(E_R(e_2)\right)^2\right\}\right]$$

$$= \left(\frac{1}{\gamma} - \frac{1}{N}\right)\frac{1}{(\gamma - 1)}\left[\sum_{i \in u}\left\{\phi_w + y_i^2\right\} - \gamma\left\{\frac{\phi_w}{\gamma} - \left(\frac{\sum_{i \in u} y_i}{\gamma}\right)^2\right\}\right]$$

$$= \left(\frac{1}{\gamma} - \frac{1}{N}\right)\phi_w + \left(\frac{1}{\gamma} - \frac{1}{N}\right)\frac{1}{(\gamma - 1)}\sum_{i \in u}\left(y_i - \frac{\sum_{i \in u} y_i}{\gamma}\right)^2.$$

So,

$$E_p E_R\left[\left(\frac{1}{\gamma} - \frac{1}{N}\right)\frac{1}{(\gamma - 1)}\sum_{i \in u}(r_i - e_2)^2 \middle| n\right]$$

$$= \left(\frac{1}{\gamma} - \frac{1}{N}\right)\phi_w + E_p\left[\left(\frac{1}{\gamma} - \frac{1}{N}\right)\frac{1}{(\gamma - 1)}\sum_{i \in u}\left(y_i - \frac{\sum_{i \in u} y_i}{\gamma}\right)^2 \middle| n\right]$$

$$= \left(\frac{1}{\gamma} - \frac{1}{N}\right)\phi_w + \left(\frac{1}{\gamma} - \frac{1}{N}\right)\frac{N\theta(1 - \theta)}{N - 1} = V(e_2) - \frac{\phi_w}{N}.$$

So,

$$\hat{V}(e_2) = \left(\frac{1}{\gamma} - \frac{1}{N}\right)\frac{1}{(\gamma - 1)}\sum_{i \in u}(r_i - e_2)^2 + \frac{\phi_w}{N},$$

is an unbiased estimator of $V(e_2)$. Proceeding similarly,

$$\hat{V}(e_3) = \left(\frac{1}{\gamma} - \frac{1}{N}\right)\frac{1}{(\gamma - 1)}\sum_{i \in u}(\hat{y}_i - e_3)^2 + \frac{\phi_w}{N\gamma}\left(\sum_{i \in u}\frac{1}{f_{is}}\right),$$

is an unbiased estimator for $V(e_3)$.

3

Randomized Response Technique in General Sampling Design

Introduction

We start with a survey population of a known number of N people labeled on identification by $i = 1, \ldots, N$, the universe being denoted by $U = (1, \ldots, i, \ldots, N)$. From this U, a sample s of labeled persons is supposed to be selected with a probability $p(s)$ according to a probability design p. For this p, the inclusion probability of an i is $\pi_i = \sum_{s \ni i} p(s)$, $i \in U$ and that of a distinct pair of individuals i, j $(i \neq j)$ is $\pi_{ij} = \sum_{s \ni i,j} p(s)$. We restrict to designs for which $\pi_i > 0 \; \forall i \in U$ and $\pi_{ij} > 0 \; \forall i, j \in U, i \neq j$. Letting y denote a real-valued variable with values y_i for $i \in U$, in sample surveys, the major problem is to estimate the total $Y = \sum_i^N y_i$ and the mean $\bar{Y} = Y/N$ on surveying a sample s of individuals bearing the values y_i for i in s. In this book we are mostly concerned with the situation when

$$y_i = 1 \quad \text{if } i \text{ bears a stigmatizing feature, say } A$$
$$= 0 \quad \text{if } i \text{ bears the complementary feature, } A^c.$$

In this case, we denote $\bar{Y}$ by θ which is the proportion of people bearing the sensitive characteristic A in a given community of N people. Our objective is to appropriately estimate this proportion θ.

Secondly, y may refer to cost unethically incurred or money unduly earned by a person i as y_i and our objective is to suitably estimate Y. In either case, if for every unit i of s the value of y_i can be directly ascertained by a direct query then a popular estimator that is often employed for Y is given by Horvitz and Thompson (1952) as

$$t = \sum_{i \in s} \frac{y_i}{\pi_i} = \sum_l \frac{y_i}{\pi_i} I_{si}.$$

Here

$$I_{si} = 1 \quad \text{if } i \in s$$
$$= 0 \quad \text{if } i \notin s.$$

Its variance is

$$V_p(t) = \sum_{i<}^{N} \sum_{j}^{(N)} \left(\frac{y_i}{\pi_i} - \frac{y_j}{\pi_j} \right)^2 (\pi_i \pi_j - \pi_{ij}) + \sum_{1}^{N} \frac{y_i^2}{\pi_i} \beta_i.$$

Here

$$\beta_i = 1 + \frac{1}{\pi_i} \sum_{j \neq i} \pi_{ij} - \sum_{1}^{N} \pi_i.$$

It is well known that if every sample s contains a common number of distinct units in it, then $\beta_i = 0$ because if that number of units called the sample size be n, then $\sum_{j \neq i} \pi_{ij}$ equals $(n-1)\pi_i$.

An unbiased estimator for $V(t)$ is

$$v_p(t) = \sum_{i<}^{N} \sum_{j}^{N} \left(\frac{y_i}{\pi_i} - \frac{y_j}{\pi_j} \right)^2 (\pi_i \pi_j - \pi_{ij}) \frac{I_{sij}}{\pi_{ij}} + \sum_{1}^{N} \frac{y_i^2}{\pi_i} \beta_i \frac{I_{si}}{\pi_i}.$$

The subscript p is used to emphasize that t is based on a sample chosen according to design p.

We shall now suppose that direct response (DR) as y_i is hard to come from the sample of individuals i in s. As a standard alternative, a suitable randomized response (RR) is gathered from every sampled i in s as a number z_i and a transform of it is derived as a number r_i with the following properties. A technique is so employed to elicit an RR from each i in s such that writing E_R, V_R to denote expectation, variance generically for any RR technique that $r_i s$ are independent, $E_R(r_i) = y_i$ for every i in U, $V_R(r_i) = V_i$ is either known for a given RRT in terms of the latter's parameter(s) or it admits an unbiased estimator v_i such that $E_R(v_i) = V_i$, $i \in U$.

It then follows that for $E = E_p$, $E_R = E_R E_p$, assuming that E_p commutes with E_R and

$$V = E_p V_R + V_p E_R = E_R V_p + V_R E_p,$$

we may observe

$$e = \sum_{i \in s} \frac{r_i}{\pi_i} = \sum_{i=1}^{N} \frac{r_i}{\pi_i} I_{si},$$

has $E(e) = E_p[E_R(e)] = E_p(t) = Y$ and also

$$E(e) = E_R \left(\sum_{1}^{N} r_i \right) = Y.$$

We then say that e is unbiased for Y. For simplicity, we shall write $\underline{Y} = (y_1, \ldots, y_i, \ldots, Y_N)$, $\underline{R} = (r_1, \ldots, r_i, \ldots, r_N)$, $R = \sum_{1}^{N} r_i$. Then it follows as in multistage sampling, that

$$V(e) = V_p(t) + \sum \frac{V_i}{\pi_i}$$

and

$$\hat{V}(e) = v_p(t)\Big|_{\underline{Y}=\underline{R}} + \sum_{i \in s} \frac{v_i}{\pi_i};$$

in case V_i is known, v_i is to be replaced by V_i.

More generally, instead of t and e above, we may employ $t_b = \sum_{i \in s} y_i b_{si}$ with b_{si} as constant, free of $\underline{Y}$, such that $\sum_{s \ni i} p(s) b_{si} = 1 \ \forall i \in U$ implying that $E_p(t_b) = Y$. Also, for $e_b = \sum_{i \in s} r_i b_{si}$, we have $E(e_b) = E_p(t_b) = Y$ and $E(e_b) = E_R(\sum_1^N r_i) = Y$ so that e_b is accepted as unbiased for Y. Further,

$$V_p(t_b) = -\sum_{i<}^{N}\sum_{j}^{N} X_i X_j \left(\frac{y_i}{X_i} - \frac{y_j}{X_j}\right)^2 d_{ij} + \sum \frac{y_i^2}{X_i}\alpha_i.$$

Writing $X_i(\neq 0)$ for any real numbers,

$$d_{ij} = E_p(b_{si} - 1)(b_{sj} - 1), \quad \alpha_i = \sum_{j=1}^{N} d_{ij} X_j,$$

and

$$\gamma_p(t_b) = -\sum_{i<}\sum_{j} X_i X_j \left(\frac{y_i}{X_i} - \frac{y_j}{X_j}\right)^2 d_{ij} \frac{I_{sij}}{\pi_{ij}} + \sum \frac{y_i^2}{X_i}\alpha_i \frac{I_{si}}{\pi_i}.$$

Then, $V(e_b) = E_R V_p(t_b) + \sum V_i$.
Here,

$$I_{sij} = I_{si} I_{sj} = 1 \text{ if } i, j \text{ both are in } s$$
$$= 0, \text{ otherwise.}$$

Obviously, $\pi_{ij} = E_p(I_{sij})$.

Then, $\quad v(e_b) = v_p(t_b)\big|_{\underline{Y}=\underline{R}} + \sum b_{si} I_{si} v_i \quad$ satisfies $\quad Ev(e_b) = V(e_b) = E_R V_p(e_b)$
$V_R E_p(e_b)$.

Alternatively,

$$V(e_b) = E_p \sum b_{si}^2 I_{si} V_i + V_p(t_b)$$
$$= E_p V_R(e_b) + V_p(E_R(e_b))$$

and

$$v'(e_b) = v_p(t_b)\big|_{\underline{Y}=\underline{R}} + \sum_{i<}\sum_{j} \frac{d_{ij}}{\pi_{ij}} I_{sij} X_i X_j \left(\frac{v_i}{X_i^2} + \frac{v_j}{X_j^2}\right)$$

is unbiased for $V(e_b)$.

To derive the above results, one may refer to Chaudhuri and Stenger (2005), pp. 15–18 and 177–179.

At this stage, it behoves us to pay attention to the specific RRTs of our interest.

Warner's Model

A sampled person i is offered a box with a number of identical cards except for a proportion $p(0 < p \neq (1/2) < 1)$ of them marked A and the rest marked A^c. On request, an RR is to be procured as

$\quad I_i = 1$ if the card drawn on request from the box before being returned unnoticed by the investigator has the type "matching" the person's true feature A or A^c.
$\quad = 0$ if there is a "mis-match."

Then,

$$E_R(I_i) = py_i + (1-p)(1-y_i)$$
$$= (1-p) + (2p-1)y_i,$$

leading to

$$r_i = \frac{I_i - (1-p)}{(2p-1)} \text{ with } E_R(r_i) = y_i, i \in U.$$

Also, r_i's are independent and

$$V_i = V(r_i) = \frac{V_R(I_i)}{(2p-1)^2} = \frac{p(1-p)}{(2p-1)^2} = \phi_p,$$

say because, $I_i^2 = I_i$ and $y_i^2 = y_i \ \forall i \in U$.
Therefore, estimating θ by

$$e = \frac{1}{N}\sum_{i \in s}\frac{r_i}{\pi_i},$$

or by

$$e_b = \frac{1}{N}\sum_{i \in s}r_i b_{si},$$

becomes a simple matter. Then,

$$N^2V(e) = \sum_{i<}^{N}\sum_{j}^{N}(\pi_i\pi_j - \pi_{ij})\left(\frac{y_i}{\pi_i} - \frac{y_j}{\pi_j}\right)^2 + \sum_{1}^{N}\frac{y_i^2}{\pi_i}\beta_i + \phi_p\sum_{1}^{N}\frac{1}{\pi_i},$$

and

$$N^2 \hat{V}(e) = \sum_{i<} \sum_j \left(\pi_i \pi_j - \pi_{ij} \right) \frac{I_{sij}}{\pi_{ij}} \left(\frac{r_i}{\pi_i} - \frac{r_j}{\pi_j} \right)^2 + \sum \frac{r_i^2}{\pi_i} \beta_i \frac{I_{si}}{\pi_i} + \phi_p \sum \frac{I_{si}}{\pi_i}.$$

Also,

$$N^2 V(e_b) = -\sum_{i<}^N \sum_j^N X_i X_j \left(\frac{y_i}{X_i} - \frac{y_j}{X_j} \right)^2 d_{ij} + \phi_p \sum_1^N \frac{y_i}{X_i} \alpha_i,$$

since $y_i^2 = y_i$ and

$$N^2 v(e_b) = -\sum_{i<}^N \sum_j^N X_i X_j \left(\frac{r_i}{X_i} - \frac{r_j}{X_j} \right)^2 d_{ij} \frac{I_{sij}}{\pi_{ij}} + \sum \frac{r_i^2}{X_i} \alpha_i \frac{I_{si}}{\pi_i} + \phi_p \sum b_{si} I_{si},$$

since $y_i^2 = y_i$ and also

$$N^2 v'(e_b) = -\sum_{i<}^N \sum_j^N X_i X_j \left(\frac{r_i}{X_i} - \frac{r_j}{X_j} \right)^2 d_{ij} \frac{I_{sij}}{\pi_{ij}} + \sum \frac{r_i^2}{X_i} \alpha_i \frac{I_{si}}{\pi_i}$$
$$+ \sum_{i<} \sum_j \frac{d_{ij}}{\pi_{ij}} I_{sij} X_i X_j \left(\frac{1}{X_i^2} + \frac{1}{X_j^2} \right) \phi_p,$$

are such that

$$Ev(e_b) = V(e_b) = Ev'(e_b).$$

It may be noted that these formulae do not reduce to the well-known formula given by Warner. This is because e and e_b do not match Warner's formula, which gives $\bar{r} = 1/n \sum_{k=1}^n r_k$; here r_k is the value of r_i for the unit chosen on the kth draw no matter how many times the same person appears in the sample giving an RR independently employing Warner's RRT each time. Naturally, $\bar{r}$ has a variance and an unbiased variance-estimator formulae not matching the above $V(e), V(e_b), \hat{V}(e), v(e_b)$, and $v'(e_b)$.

Unrelated Question Model

Horvitz et al. (1967) followed by Greenberg et al. (1969) recommend an alternative to Warner's (1965) RRT. They suggest that like A, the attribute A^c may be stigmatizing as well. So, people may be uncomfortable to give out their features A and A^c, even though by a randomized procedure without divulging to the investigator the outcome actually observed on the RRT. Along with the sensitive attribute A say, "habitual drunken driving" under investigation they conceptualize an apparently unrelated one, say, B, like "preferring" "music to painting" or "football to tennis," and so on. So, while treating y to denote bearing A, another variable x is introduced to denote bearing B rather than its complement A^c, say.

Thus,

$$y_i = 1 \quad \text{if } i \text{ bears } A$$
$$= 0 \quad \text{if } i \text{ bears } A^c$$

and

$$x_i = 1 \quad \text{if } i \text{ bears } B$$
$$= 0 \quad \text{if } i \text{ bears } B^c.$$

The objective as before is to estimate $\theta = 1/N \sum_1^N y_i$.

The RRT propagated by Horvitz et al. (1967) and Greenberg et al. (1969) is called "unrelated question model" and is employed as follows. Two boxes are filled with a large number of similar cards except that in the first box a proportion $p_1(0 < p_1 < 1)$ of them is marked A and the complementary proportion $(1 - p_1)$ each bearing the mark B, while in the second box these proportions are p_2: $(1 - p_2)$ keeping p_2 different from p_1. In their original RRT, two independent SRSWRs of sizes n_1 and n_2 are taken, in one, the first box and in the other, the second box is offered to each sample person whenever addressed. But we prefer to deviate from this retaining the essential features of their RRT in the following manner.

In our modification, a single sample is chosen and every person sampled is requested to draw one card randomly from the first box and repeat this independently with the second box and in the first case give an RR as

$I_i = 1$ if the card type "matches" person's A or B characteristic
$\quad = 0$ if it is a "No match"

and in the second case give an RR as

$J_i = 1$ if there is a "Match"
$\quad = 0$ if there is a "No Match,"

whoever may be the person labeled i as sampled.

It follows that

$$E_R \left(I_i \right) = p_1 y_i + \left(1 - p_1 \right) x_i$$

and

$$E_R \left(J_i \right) = p_2 y_i + \left(1 - p_2 \right) x_i,$$

giving us

$$E_R\left[(1-p_2)I_i - (1-p_1)J_i\right] = (p_1 - p_2)y_i,$$

leading to

$$r_i = \frac{(1-p_2)I_i - (1-p_1)J_i}{(p_1 - p_2)} \text{ such that } E_R(r_i) = y_i.$$

Since $y_i^2 = y_i$, $x_i^2 = x_i$, $I_i^2 = I_i$ and $J_i^2 = J_i$, we get

$$V_R(I_i) = E_R(I_i)(1 - E_R(I_i)) = p_1(1 - p_1)(y_i - x_i)^2.$$

Similarly, $V_R(J_i) = p_2(1 - p_2)(y_i - x_i)^2$

$$V_R(r_i) = \frac{(1-p_2)^2 V_R(I_i) + (1-p_1)^2 V_R(J_i)}{(p_1 - p_2)^2}$$

$$= \frac{(1-p_1)(1-p_2)(p_1 + p_2 - 2p_1p_2)}{(p_1 - p_2)^2}(y_i - x_i)^2$$

As $V_R(r_i) = V_i$ is known only in terms of the parameters p_1, p_2 of the RRT but involves unknown quantities $(y_i - x_i)^2$, it is hard to visualize possible elegance in the extension presented in case of Warner's (1965) RRT when linked to SRSWR or inverse SRSWR utilizing distinct units with single or multiple RRs. However, unbiased estimation of V_i is a simple matter as shown below:

$$V_i = V_R(r_i) = E_R(r_i^2) - \left[E_R(r_i)\right]^2$$
$$= E_R(r_i^2) - y_i^2 = E_R(r_i^2) - y_i$$
$$= E_R(r_i^2) - E_R(r_i) = E_R\left[r_i(r_i - 1)\right].$$

So, $r_i(r_i - 1) = \hat{V}_i$ may be taken as an unbiased estimator for V_i.
Therefore, estimation of θ by

$$e = \frac{1}{N}\sum_{i\in s}\frac{r_i}{\pi_i} \text{ or by } e_b = \frac{1}{N}\sum r_i b_{si} I_{si},$$

is a simple matter as before and we need not repeat the algebraic details here as they are already covered in this chapter under "Warner's Model."

An insurmountable problem here is that "$r_i(r_i - 1)$" may turn out negative though unbiased for a nonnegative number, namely V_i.

Comparison of $V_i = \phi_w = [p(1-p)/(2p-1)^2]$ for Warner's RRT versus

$$V_i = \frac{p_1(1-p_1)(p_1 + p_2 - 2p_1 p_2)}{(p_1 - p_2)^2} - (y_i - x_i)^2,$$

is also virtually impossible. So, relative performances of $V(e)$ or $V(e_b)$ for RRT versus "unrelated question model" is rather impossible. However, performances as measured by coefficients of variation for estimators of θ based on various RRTs is not difficult, as we shall illustrate later for a few cases.

Kuk's RRT

A sampled person i is to be approached with two boxes with cards having proportions θ_1 and θ_2, respectively ($0 < \theta_1 \neq \theta_2 < 1$) that are "red" in color, the remaining cards being "nonred." A sampled person is to draw $K(>1)$ cards with replacement and report the number out of the K cards drawn that are "red," without disclosing to the investigator which box is used to draw. The understanding is that if the person i bears A, the first box is to be used and if the person bears A^c, the second box is to be chosen.

Let for the ith person sampled, the RR be f_i which is the number of red cards drawn out of the $K(>1)$ independent trials made with replacement. Since $y_i = 1$ if i bears A and $y_i = 0$ if i bears A^c the f_i being "binomially distributed," one has

$$E_R(f_i) = K\left[y_i \theta_1 + (1 - y_i)\theta_2\right]$$
$$= K\left[\theta_2 + y_i(\theta_1 - \theta_2)\right],$$

and

$$V_R(f_i) = K\left[y_i \theta_1 (1 - \theta_1) + (1 - y_i)\theta_2 (1 - \theta_2)\right]$$
$$= K\left[\theta_2 (1 - \theta_2) + y_i \{\theta_1 (1 - \theta_1) - \theta_2 (1 - \theta_2)\}\right].$$

Then writing

$$r_i(K) = \frac{\left(\dfrac{f_i}{K} - \theta_2\right)}{(\theta_1 - \theta_2)} \quad \text{one has } E_R\left(r_i(K)\right) = y_i,$$

and

$$V_R\left(r_i(K)\right) = \frac{V_R\left(\dfrac{f_i}{K}\right)}{\left(\theta_1 - \theta_2\right)^2} = V_i(K),$$

say,

$$= b_i(K)y_i + C_i(K),$$

writing

$$b_i(K) = \frac{(1-\theta_1-\theta_2)}{K^2(\theta_1-\theta_2)^2}, \quad C_i(K) = \frac{\theta_2(1-\theta_2)}{K^2(\theta_1-\theta_2)^2}.$$

Then, $v_i(K) = b_i(K)r_i(K) + C_i(K)$ is an unbiased estimator for $V_i(K)$ because $E_R v_i(K) = b_i(K)y_i + C_i(K)$. If we decide on a prior choice of $K(>1)$ it will be better to suppress K throughout the parentheses above. Now the theory needed to estimate $\theta = 1/N \sum_1^N y_i$ is obvious.

We should mention that the above description essentially modifies Kuk's (1990) original narration as it was related to the SRSWR of the respondents.

Let us briefly take into account his theory. With Kuk's (1990) RRT as described, when a person i is selected by SRSWR in n draws, on suppressing $K(>1)$ in the parentheses in $r_i(K)$ we get

$$E_R\left(\frac{f_i}{K}\right) = \theta\,\theta_1 + (1-\theta)\theta_2 = \phi, \text{ say } \forall i$$

Writing $\bar{r} = (1/n)\sum'(f_i/K), \sum'$ the sum over the number of times ith person appears in the sample, an unbiased estimator for θ follows as $\hat{\theta} = \bar{r} - \theta_2/\theta_1 - \theta_2$. Naturally, one derives

$$V\left(\hat{\theta}\right) = \frac{1}{n(\theta_1-\theta_2)^2}\left[\frac{\phi(1-\phi)}{K} + \theta(1-\theta)(\theta_1-\theta_2)^2\left(1-\frac{1}{K}\right)\right]$$

Kuk, however, did not supply any unbiased estimator for $V(\hat{\theta})$. Let us provide one below on accomplishing the following analyses:

$$V_R\left(\frac{f_i}{K}\right) = \frac{\phi(1-\phi)}{K} + \theta(1-\theta)(\theta_1-\theta_2)^2\left(1-\frac{1}{K}\right) = W, \text{ say } \forall i \in U$$

Let $R_{ij} = 1$ if the ith person gets a "red" card on the jth draw,
$\quad = 0$ if on the jth draw, i gets a "nonred" card.

Letting

$$
w_i = \frac{1}{K(K-1)} \left[\sum_{j=1}^{K} R_{ij} - \frac{\sum_{j=1}^{K} R_{ij}}{K} \right]^2
$$

$$
= \frac{1}{K(K-1)} \left[\sum_{1}^{K} R_{ij} - \frac{\left(\sum_{1}^{K} R_{ij}\right)^2}{K} \right],
$$

it follows that $E_R(w_i) = W, \forall i \in U$.

So, $v = 1/[n^2(\theta_1 - \theta_2)^2]\Sigma' w_i$ is an unbiased estimator for $V(\hat{\theta})$.

Because of the complicated form of $V_R(f_i/K) = W$ which involves unknown parameters θ and ϕ it does not seem analytically worthwhile to extend our works relating to distinct units with single or multiple application of Warner's RRT in SRSWR in n draws or inverse SRSWR to the case of Kuk's (1990) RRT.

This need also does not seem compelling as the theory applicable to any sampling design for Kuk's RRT is already clarified as above, and as usual using

$$
e = \sum_{i \in s} \frac{r_i}{\pi_i}, \quad e_b = \sum_{i \in s} r_i b_{si}.
$$

Here of course s is a sample chosen according to a suitable design p, $\pi_i = \Sigma_{s \ni i}\, p(s)$, assumed positive, $\forall i \in U$ and b_{si} are constants free of $\underline{Y} = (y_1, \ldots, y_i, \ldots, y_N), \underline{R} = (r_1, \ldots, r_i, \ldots, r_N)$.

Christofides's RRT

A sampled person i is to be approached with a box of $M(\geq 2)$ cards of common shape, size, weight, color, height, breadth, and thickness but bearing different marks $1, \ldots, j, \ldots, M$ in proportions $p_1, \ldots, p_j, \ldots, p_M$ $(0 < p_j < 1, \Sigma_{j=1}^{M} p_j = 1)$ with a request to draw randomly one of these cards. The RR to be reported is K if the person's attribute is A^c or $(M + 1 - K)$ if it is A; of course the true attribute is to be hidden from the investigator and the actual mark on the card happens to be K.

So, we may represent the RR available from the ith person as

$$
z_i = (M + 1 - K)y_i + K(1 - y_i), \quad K = 1, \ldots, M.
$$

Since $E_R(K) = \sum_{K=1}^{M} K p_K = \mu$, say, and $V_R(K) = \sum_{K=1}^{M} K^2 p_K - \mu^2$ it follows that $E_R(z_i) = y_i(M + 1 - 2\mu) + \mu = \mu_i$, say, and

$$
\begin{aligned}
V_R(z_i) &= E_R(z_i^2) - \mu_i^2 \\
&= E_R\left[(M + 1 - K)^2 y_i + K^2(1 - y_i)\right] - \mu_i^2,
\end{aligned}
$$

because $y_i^2 = y_i$

$$
\begin{aligned}
&= \sum_{K=1}^{M} p_K K^2 + y_i \sum_{K=1}^{M} p_K\left[(M + 1 - K)^2 - K^2\right] - \mu_i^2 \\
&= \sum p_K K^2 + y_i \sum p_K\left[(M + 1)(M + 1 - 2K)\right] \\
&\quad - \mu^2 - y_i(M + 1 - 2\mu)^2 - 2y_i(M + 1 - 2\mu)\mu \\
&= \left(\sum p_K K^2 - \mu^2\right) + y_i\left[(M + 1)^2 - 2(M + 1)\mu - (M + 1)^2\right. \\
&\quad \left. - 4\mu^2 + 4(M + 1)\mu - 2(M + 1)\mu + 4\mu^2\right] \\
&= \sum p_K K^2 - \mu^2,
\end{aligned}
$$

that is,

$$
V_R(z_i) = V_R(K) = \sum p_K K^2 - \mu^2.
$$

Let us take

$$
r_i = \frac{z_i - \mu}{M + 1 - 2\mu}.
$$

Then,

$$
E_R(r_i) = y_i,
$$

and

$$
V_R(r_i) = \frac{V_R(z_i)}{(M + 1 - 2\mu)^2} = \frac{V_R(K)}{(M + 1 - 2\mu)^2} = V_i, \text{ say.}
$$

If M is taken as $= $ to 2, and write $p_1 = p$ so that $p_2 = 1 - p$, then one gets $z_i = K + (3 - 2K)y_i$, $K = 1, 2$, $M = 2$, $\mu_i = 2 - p$,

$$
r_i = \frac{z_i - (2 - p)}{2p - 1} \quad \text{and} \quad V_R(r_i) = \frac{p(1 - p)}{(2p - 1)^2} = \phi_w.
$$

Needless to mention, in this case Christofides's (2003) RRT is relegated to Warner's RRT as it should be.

What we have described above is of course our (cf. Chaudhuri, 2004) version of the original RRT of Christofides (2003) needlessly tied to SRSWR as are most of the RRTs. In his original scheme using the box of cards numbered $1, \ldots K, \ldots, M$ in proportions $p_1, \ldots, p_K, \ldots, p_M$ with $0 < p_K < 1$, $K = 1, \ldots, M$, $\sum_1^M p_K = 1$, a person labeled i chooses a card marked K with probability p_K, $K = 1, \ldots, M$ and reports z_i as $(M + 1 - K)$ if the person bears A and as K if the person bears A^c. If a person i chosen by SRSWR in n draws the RR z_i has the expectation $E_R(z_i) = \mu + \theta(M + 1 - 2\mu)$ $\forall i \in U$ no matter on which draw of the SRSWR the ith person happens to be selected.

In fact,

$$E_R\left(z_i\right) = E_R\left(K\right) + \theta\left(M + 1 - E_R\left(K\right)\right) = \mu_i,$$

with

$$E_R\left(K\right) = \sum_{K=1}^{M} Kp_K = \mu \quad \text{and} \quad V_R\left(K\right) = \sum K^2 p_K - \mu^2,$$

$$
\begin{aligned}
V_R\left(z_i\right) &= E_R\left(z_i^2\right) - \mu_i^2 \\
&= (1 - \theta)\sum K^2 p_K + \theta\sum (M + 1 - K)^2 p_K - \mu_i^2 \\
&= \theta(M + 1)^2 + \sum K^2 p_K - 2\theta(M + 1)\mu - \mu_i^2 \\
&= \theta(M + 1)^2 + \sum K^2 p_K - 2\theta(M + 1)\mu - \mu^2 - \theta^2 (M + 1 - \mu)^2 \\
&\quad - 2\theta(M + 1 - \mu)\mu \\
&= V_R\left(K\right) + \theta(1 - \theta)(M + 1 - 2\mu)^2.
\end{aligned}
$$

Let $\bar{z} = (1/n)\sum' z_i$, the sum being over the draws in which the unit labeled i happens to appear in the SRSWR in n draws and reports the RR as K or $(M + 1 - K)$.

Then, $E_R(\bar{z}) = \mu + \theta(M + 1 - 2\mu)$. So, an unbiased estimator for θ is $\hat{\theta} = [(\bar{z} - \mu)/(M + 1 - 2\mu)]$.

Then,

$$
\begin{aligned}
V(\hat{\theta}) &= \frac{V(\bar{z})}{(M + 1 - 2\mu)^2} \\
&= \frac{1}{n} \frac{1}{(M + 1 - 2\mu)^2} V(z_i) \\
&= \frac{1}{n} \frac{1}{(M + 1 - 2\mu)^2}\left[\left(\sum K^2 p_k - \mu^2\right)\right] + \frac{\theta(1 - \theta)}{n},
\end{aligned}
$$

because $V(\bar{z}) = (1/n)V_R(z_i)$ for $i \in U$.

To avoid complicated algebra, we do not study alternative estimators competitive with $\hat{\theta}$ above to take into account distinct units with single or multiple RRs using Christofides's RRT and extension to inverse SRSWR. This may also be overlooked as we have covered the general case with unequal probability sampling.

Forced Response Scheme

For a person labeled i in $U = (1, ..., i, ..., N)$ y_i is the number 1 if i bears the sensitive attribute A or the number 0 if the person bears the complementary attribute A^c. To estimate $\theta = 1/N \sum_1^N y_i$ the forced response technique is applied in the following manner, no matter how selected, the person labeled i is approached with a box containing a large number of cards marked "yes," "no" and "genuine" in proportions p_1, p_2, and $(1 - p_1 - p_2)$, respectively $(0 < p_1, p_2 < 1, p_1 + p_2 < 1)$. The person is requested to draw one card after shaking the box thoroughly and not letting the investigator know the type of card drawn and is requested to report yes or no if the card type is marked "yes" or "no" or genuinely as yes or no if a card marked "genuine" is drawn and, if the person bears A or A^c, respectively. Thus, the RR from i is marked

$$I_i = 1 \quad \text{if } i \text{ responds "yes"}$$
$$= 0 \quad \text{if } i \text{ responds "no."}$$

Then we have the following results:

$$Prob \ (I_i = 1 | y_i = 1) = p_1 + (1 - p_1 - p_2) = 1 - p_2,$$
$$Prob \ (I_i = 0 | y_i = 1) = p_2, \quad Prob \ (I_i = 1 | y_i = 0) = p_1,$$
$$Prob \ (I_i = 0 | y_i = 0) = p_2 + (1 - p_1 - p_2) = 1 - p_1.$$

So, $E_R(I_i) = Prob(I_i = 1) = p_1 + y_i(1 - p_1 - p_2).$
Then, let us define

$$r_i = \frac{I_i - p_1}{(1 - p_1 - p_2)}. \quad \text{Then,} \quad E_R(r_i) = y_i.$$

Therefore,

$$V_R(r_i) = V_i = \frac{V_R(I_i)}{(1 - p_1 - p_2)^2} = \frac{E_R(I_i)(1 - E_R(I_i))}{(1 - p_1 - p_2)^2}$$

$$= \frac{p_1(1 - p_1) + y_i(1 - p_1 - p_2)(p_2 - p_1)}{(1 - p_1 - p_2)^2}.$$

Hence,

$$V_i = \frac{p_1(1-p_1)}{(1-p_1-p_2)^2}, \quad \text{if} \quad y_i = 0$$
$$= \frac{p_2(1-p_2)}{(1-p_1-p_2)^2}, \quad \text{if} \quad y_i = 1.$$

Chaudhuri and Mukerjee (1988, pp. 16–17) described this scheme applicable in an SRSWR in n draws. In this case, as discussed also by Fox and Tracy (1986)

$$Prob(\text{Yes Response}) = Prob(I_i = 1)$$
$$= p_1\theta + p_2$$

whichever may be the draw in which the ith person happens to be selected. So,

$$\hat{\theta} = \frac{(1/n)\sum I_I - p_2}{p_1},$$

satisfies $E(\hat{\theta}) = \theta$.

Since $Prob(I_i = 1) = E_R(I_i) = p_i\theta + p_2 = \lambda$, say and $V_R(I_i) = \lambda(1-\lambda) = E_R(I_i)$ $(1 - E_R(I_i))$, we may readily say that $\sum'I_i$ has the binomial distribution $b(.; n, \lambda)$ and hence

$$V(\hat{\theta}) = \frac{\lambda(1-\lambda)}{n} = \frac{(p_1\theta + p_2)(1 - p_2 - p_1\theta)}{n},$$

and an unbiased estimator for

$$V(\hat{\theta}) \text{ is } \hat{V}(\hat{\theta}) = \frac{\hat{\lambda}(1-\hat{\lambda})}{(n-1)},$$

with $\hat{\lambda} = p_1\hat{\theta} + p_2$.

An extension of this forced response technique to the case of distinct units with single or multiple responses in an SRSWR with a fixed number of n draws or in inverse SRSWR is quite feasible. But we do not pursue as the algebra is not quite elegant and the case of extension to general unequal probability sampling schemes is obviously clarified above as r_i with $E_R(r_i) = y_i$ and $V_R(r_i) = V_i$ are already specified. Clearly,

$$\hat{V}_i = \frac{p_1(1 - p_1) + r_i(1 - p_1 - p_2)(p_2 - p_1)}{(1 - p_1 - p_2)^2},$$

is an unbiased estimator for V_i.

Mangat and Singh's RRT

This RRT is a "slight modification" of Warner's (1965) RRT in the following sense. A sampled person i is presented with two boxes. The person is requested first to randomly draw one card out of a number of cards marked T or R in proportions $T: (1 - T)$, $(0 < T < 1)$. If a T marked card is drawn, the person is to report the true value y_i, which as usual is 1 if i bears A and 0 if i bears A^c. If an R marked card is drawn the person is to draw one card from the second box containing cards marked A and A^c in proportions $p: (1 - p), (0 < p \neq 1/2 < 1)$. The person's RR then will be

$$\begin{aligned} I_i &= 1 \quad \text{if the card type matches the true feature } A \text{ or } A^c \\ &= 0 \quad \text{if there is "no match."} \end{aligned}$$

The actual operation is not to be disclosed to the investigator who is to only observe the number reported by i. The RR from i may be denoted as z_i. Then,

$$\begin{aligned} z_i &= y_i \quad \text{if the first box gives a } T\text{-card} \\ &= I_i \quad \text{if the first box gives an } R\text{-card.} \end{aligned}$$

Then, $Prob(z_i = y_i) = T + (1 - T)p = 1 - \alpha$, say, and $Prob(z_i = 1 - y_i) = (1 - T)(1 - p)$ and obviously, $(1 - T)(1 - p)$ equals α. So,

$$\begin{aligned} E_R(z_i) &= Ty_i + (1 - T)\big[py_i + (1 - p)(1 - y_i)\big] \\ &= (1 - T)(1 - p) + y_i\big[T + (1 - T)(2p - 1)\big]. \end{aligned}$$

Therefore,

$$r_i = \frac{z_i - (1 - T)(1 - p)}{T + (1 - T)(2p - 1)},$$

has $E_R(r_i) = y_i$ on ensuring the choice $T + (1 - T)(2p - 1) \neq 0$

$$E_R(z_i) = (1 - T)(1 - p) + y_i\left[T + (1 - T)(2p - 1)\right]$$
$$= (1 - T)(1 - p) + y_i\left[(T + 1 - T)p - (1 - T)(1 - p)\right]$$
$$= \alpha + y_i(1 - 2\alpha),$$

because
$$\alpha = (1 - T)(1 - p) \quad \text{and} \quad 1 - \alpha = T + (1 - T)p.$$

Now,
$$V_R(z_i) = E_R(z_i)\left(1 - E_R(z_i)\right),$$

because $z_i^2 = z_i$.
So,

$$V_R(z_i) = \left[\alpha + y_i(1 - 2\alpha)\right]\left[(1 - \alpha) - y_i(1 - 2\alpha)\right] = \alpha(1 - \alpha),$$

because $y_i^2 = y_i$ and the coefficient of y_i is zero.
So,

$$V_R(z_i) = (1 - T)(1 - p)\left[T + (1 - T)p\right]$$
$$= (1 - T)(1 - p)\left[T + p - pT\right].$$

So,

$$V_i = V_R(r_i) = \frac{(1 - T)(1 - p)\left[T + p(1 - T)\right]}{\left[T + (1 - T)(2p - 1)\right]^2}.$$

If $T = 0$, V_i reduces to

$$\frac{p(1 - p)}{(2p - 1)^2} = \phi_w.$$

So, a further study of Mangat and Singh's (1990) RRT in estimating $\theta = 1/N\sum_1^N y_i$ is simple enough as with Warner's RRT in case the respondents are sampled with general unequal probability sampling designs.

Mangat's Scheme

Mangat (1992), as a matter of fact, essentially applied the same amendment to the unrelated question (URL) model of Horvitz et al. (1967) and Greenberg et al. (1969) as was applied to Warner's (1965) model by Mangat and Singh (1990). For simplicity, he considered the situation when the proportion of people π_B bearing an innocuous characteristic B as known. His RRT enjoins a sampled person to draw randomly a card from a box with a proportion T $(0 < T < 1)$ of cards marked T and the rest marked R. The instruction for the person is to say "yes" if the person draws a T-marked card and bears the stigmatizing feature A. If an R marked card is drawn, the person is instructed to use a second box with cards marked A or B in proportions p: $(1 - p)$, $(0 < p < 1)$ and report "yes" or "no" if the card type matches or does not match the actual feature. The box and the card type used should not be disclosed. But the response must be truthful. For simplicity, SRSWR in n draws is assumed. Since two boxes are needed for this RR device, Mangat (1992) calls it a "two-stage" RRT. With θ, ψ, and λ, respectively, as the probability that a person bears A, B and responds "yes," he observes

$$\lambda = T\theta + (1 - T)\big[p\theta + (1 - p)\psi\big]$$
$$= \big[T + p(1 - T)\big]\theta + (1 - T)(1 - p)\psi.$$

With n' as the observed frequency of "yes" the rest of the theory is easy and we choose to skip.

Our plan here is to extend this RRT of Mangat (1992) to cover the case when a general sampling is employed and ψ, the proportion of people bearing the innocuous characteristic B is unknown. Then the device needs three boxes—the first one is exactly as before, the second one is same as before but with p replaced by p_1 and the third box contains a proportion p_2 $(0 < p_2 < 1$ but $p_1 \neq p_2)$ of cards marked A and the rest marked B. The instruction to a sampled respondent labeled i is as before to tell the truth first using "the first box and if necessary also the second box" and next independently make a second truthful response using similarly "the first box and if necessary, the third box."

Let

$$\begin{aligned}
y_i &= 1/0 \quad \text{if } i \text{ bears } A \text{ or not,} \\
x_i &= 1/0 \quad \text{if } i \text{ bears } B \text{ or not,} \\
I_i &= 1/0 \quad \text{if RR is "yes"/"no" for the first trial,} \\
J_i &= 1/0 \quad \text{if RR is "yes"/"no" for the second trial if executed} \\
&\qquad \text{by the } i\text{th person sampled.}
\end{aligned}$$

Then,

$$E_R(I_i) = Ty_i + (1 - T)[p_1 y_i + (1 - p_1)x_i]$$
$$E_R(J_i) = Ty_i + (1 - T)[p_2 y_i + (1 - p_2)x_i] \quad \text{leading to}$$

$$r_i = \frac{(1 - p_2)I_i - (1 - p_1)J_i}{(p_1 - p_2)},$$

$E_R(r_i) = y_i$ and since $y_i^2 = y_i$, $V_R(r_i) = V_i$ say, equals $E_R[r_i(r_i - 1)]$ and hence $r_i(r_i - 1) = \hat{V}_i$ is an unbiased estimator for $V_R(r_i)$.

Consequently, an estimator e for θ and v for $V(e)$, the variance of e follows as usual.

Mangat, Singh, and Singh's Scheme

Mangat et al. (1992) gave an RR device again as a modification to the basic URL model but again assuming π_B as known in a manner akin to Mangat's (1992) discussed in this chapter under "Mangat's Scheme."

A person labeled i, if sampled, is offered a box and is to answer "yes" if the person bears A. But if the person bears A^c then the person is to draw a card from the box with a proportion p ($0 < p < 1$) of cards marked A and the rest marked B; if the person draws a card marked B is to say "yes" again if the person actually bears B; in any other case "no" is to be answered; the actual implementation of this device is to be concealed.

They considered exclusively SRSWR to sample the people. Consequently,

$$\lambda = Prob\left["yes"\text{response}\right] = \theta + (1 - \theta)(1 - p)\psi$$
$$= \left[1 - (1 - p)\psi\right]\theta + (1 - p)\psi.$$

Since ψ is known as a positive proper fraction, using the number of "yes" responses n^1 in a total sample in n draws the rest of the estimation theory needed follows quite easily. We choose to omit.

Instead, we consider an extension of this RRT to cover the case of a general sampling scheme and one with a natural situation with an unknown value of ψ the proportion of people bearing the innocuous characteristic B. For a person labeled i who is chosen, following a general scheme of sampling, the instruction is to say "yes" if the person bears A, and if not, to take randomly a card from a box containing a proportion p_1 ($0 < p_1 < 1$) of cards marked A and the rest marked B and to say "yes" if a B-marked card is drawn and if the person really bears B; otherwise, the response will be "no." However, this entire exercise is to be repeated independently with the only change that the proportion of A-marked cards is p_2 and $p_2 \neq p_1$, ($0 < p_2 < 1$). As before, letting

$y_i = 1/0$ if i bears A/A^c, $x^i = 1/0$ if i bears B/B^c,

$I_i = 1$ if a "yes" response emerges in the first trial, 0 otherwise;

$J_i = 1/0$ if "yes"/"no" is the RR in the second.

Then,

$$E_R\left(I_i\right) = y_i + (1 - y_i)(1 - p_1)x_i$$
$$= \left[1 - (1 - p_1)x_i\right]y_i + (1 - p_1)x_i,$$
$$E_R\left(J_i\right) = y_i + (1 - y_i)(1 - p_2)x_i$$
$$= \left[1 - (1 - p_2)x_i\right]y_i + (1 - p_2)x_i,$$

leading to

$$r_i = \frac{(1 - p_2)I_i - (1 - p_1)J_i}{(p_1 - p_2)}, \quad E_R(r_i) = y_i,$$

$V_i = V_R(r_i) = E_R[r_i(r_{i-1})]$ and hence $\hat{V}_i = r(r_i - 1)_i$ is an unbiased estimator of V_i. The problem of estimation is solved as usual. The formulae for $V_R(r_i)$ with r_i in this section as well as under "Mangat's Scheme" are not simple enough as they involve y_i, x_i for $i = 1, ..., N$; we do not consider it worthwhile to recommend further study of these RRTs with SRSWR utilizing single or multiple responses from the distinct persons sampled.

Singh and Joarder's Scheme

Singh and Joarder (1997) gave an RRT and showed that if RRs are observed from an SRSWR sample in n draws, then their estimator for θ has less of a variance than that of Warner's (1965) estimator for θ. Their RRT is as follows: A respondent bearing A^c is to respond as in Warner's RRT but one bearing A is to postpone the response to a second performance of Warner's RRT unless the first one induces a "yes" response. Of course, a pack of cards containing a proportion p $(0 < p < 1)$ marked A and the rest marked A^c is to be used without divulging anything about this actual performance of the RRT to the interviewer as usual. Naturally, for any respondent chosen in an SRSWR,

$$Prob["yes" \text{ response}] = (1 - p)(1 - \theta) + \theta[p(1 - p) + p]$$
$$= (1 - p) + \theta[(2p - 1) + p(1 - p)].$$

So, choosing $p \in (0, 1)$, such that $(2p - 1) + (1 - p) \neq 0$, it is a simple matter to estimate θ using the total number of "yes" responses in an SRSWR in n draws. Deriving the variance of the estimator of θ and an unbiased estimator of the variance is also not a problem. But the variance involves θ, the unknown element. So, we are not interested here in studying the consequences of identifying the distinct persons sampled and examining the consequences of single or multiple responses from each of them.

Instead, we consider examining the RRT of Singh and Joarder (1997) when the sampling is from a finite population with general unequal probabilities.

Letting $y_i = 1/0$ as i bears A/A^c and $I_i = 1$ if "yes" is the RR from the ith person, we have

$$E_R(I_i) = (1 - p)(1 - y_i) + y_i[p + p(1 - p)]$$
$$= y_i[(2p - 1) + p(1 - p)] + (1 - p).$$

Then, for $r_i = [I_i - (1 - p)/\{(2p - 1) + p(1 - p)\}]$, taking the denominator nonzero, we have $E_R(r_i) = y_i$. Since $y_i^2 = y_i$, $V_i = V_R(r_i) = E_R r_i(r_{i-1})$ yielding $\hat{V}_i = r_i(r_{i-1})$ as an unbiased estimator for V_i. Deriving e as an unbiased estimator for θ and v as an unbiased estimator for $V(e)$ is no more a great problem if one employs a consistent sampling scheme, with $\pi_i > 0, \pi_{ij} > 0$.

Chaudhuri and Pal (2002) had worked out

$$V_i = 1/\alpha^2[E_R(I_i)(1 - E_R)(I_i)], \alpha = (2p - 1) + p(1 - p) \text{ as}$$
$$V_i = 1/\alpha^2[\beta y_i + p(1 - p)], \beta = \alpha(1 - \alpha) - 2\alpha(1 - p).$$

Noting that α, β are known, they gave $v_i = 1/\alpha^2[\beta r_i + p(1 - p)]$ as an unbiased estimator for V_i.

In general, if for any RRT we may express $V_R(r_i)$ in the form $V_i = ay_i + b$, with a, b known, as it is almost always the case with qualitative characteristics carrying social stigma, then V_i may be clearly estimated unbiasedly by $v_i = ar_i + b$.

A few more RRT schemes are analyzed further.

Dalenius and Vitale's Scheme

Dalenius and Vitale (1974) have given an RRT that really refers to a quantitative but sensitive variable whose values are grouped into a number of disjoint classes so that a response solicited from a sampled person is to refer to one of these classes in which the respondent is placed in terms of the person concerned's value of the stigmatizing variable. The main difference of their RRT from the ones already discussed earlier is that though every person labeled i gives a response "yes" or a "no" the RR (the ith person's response) does not vary with i, rather it remains a constant for every i, in U. Estimation, however, proceeds easily both with SRSWR and general sampling schemes in very simple ways. So, it is worthwhile to examine the extension to SRSWR with single or multiple responses from the distinct persons sampled by SRSWR. No one so far is known to have ventured in this research area. To describe Dalenius and Vitale (1974) in detail, let us consider the details narrated by Chaudhuri (2002). Suppose a person has y acres of crop area spread over various local areas and naturally there is an apathy against the disclosure of the amount, which tantamounts to revealing the owner's assets. Let (a_0, a_T) denote the known interval in which y lies. Let this interval be composed of the nonoverlapping subintervals, $(a_0, a_1), \ldots, (a_{j-1}, a_j), \ldots, (a_{T-1}, a_T)$

each with an equal width h and x_j be the mid point or the class mark of the jth class $(a_{j-1}, a_j), j = 1,..., T$. Let for a suitably fixed constant $C, j = (x_j - C)/h$ and f_j be the unknown frequency of the jth class, $\sum_{j=1}^{T} f_j = F$ and $\theta_j = f_j/F$ be the corresponding relative frequency, such that $\sum_1^T \theta_j = 1$. Then, the mean of y is $\psi = C + h \sum_{j=1}^{T} j\theta_j$.

Dalenius and Vitale's (1974) RRT is to unbiasedly estimate

$$\mu = \sum_1^T j\theta_j,$$

and from that derive an estimator for ψ, the ultimate objective.

According to their RRT, a sampled person is given a disk on which T equal segments are marked by $1, ..., j, ..., T$ and is instructed to spin a "pointer" on the disk and to respond "yes" if the "pointer" stops in a segment marked "less than or equal" to the mark representing the class to which the person belongs in terms of the sensitive variable of our interest; otherwise the RR is to be "no." Writing

$$I_i = 1, \quad \text{if } i\text{th person responds "yes"}$$
$$= 0, \quad \text{if the response is "no,"}$$

then,

$$E_R(I_i) = \frac{1}{T}\sum_{j=1}^{T} j\theta_j = \frac{\mu}{T} = \forall i$$

Prob ["yes" response] $= \lambda$, say.

So, TI_i is an unbiased estimator for μ for every i in a sample.

Also,

$$V_R(I_i) = E_R(I_i)(1 - E_R(I_i))$$
$$= \frac{\mu}{T}\left(1 - \frac{\mu}{T}\right) \quad \forall i \in U$$
$$= \lambda(1 - \lambda).$$

So, $V_i = V_R(I_i)$ cannot be easily estimated unbiasedly. However, Dalenius and Vitale (1974) treated only the case of SRSWR in n draws. So, writing λ as the proportion of "yes" responses in the sample, λ unbiasedly estimates λ and hence $\hat{\mu} = T\hat{\lambda}$ unbiasedly estimates μ and

$$V(\hat{\mu}) = T^2 \frac{\lambda(1 - \lambda)}{n},$$

and hence $V(\hat{\mu})$ is unbiasedly estimated by

$$\hat{V}(\hat{\mu}) = T^2 \frac{\hat{\lambda}(1 - \hat{\lambda})}{(n - 1)}.$$

For the general sampling case, an estimator e works out easily but not v. So, we need a second RR independently of I_i from every sampled person adopting the same RRT, denoted as

$$I'_i = 1 \quad \text{if "yes" is the response}$$
$$= 0 \quad \text{if "no" is the response.}$$

Then, $E_R(I'_i) = E_R(I_i) = \mu/T$.

So, $J_i = 1/2(I_i + I'_i)$ satisfies $E_R(J_i) = \mu/T \Rightarrow E_R(TJ_i) = \mu$ that is, TJ_i is unbiased for μ and e is to be derived using TJ_i for $i \in s$. Now, $V_i = V_R(J_i) = 1/4[V_R(I_i) + V_R(I'_i)]$ is unbiasedly estimated by $\hat{V}_i = 1/4(I_i - I'_i)^2$ and so $v_i = T^2/4(I_i - I'_i)^2$ unbiasedly estimates $V_R(TJ_i)$. So for $\hat{\mu} = TJ_i$ an unbiased variance estimator is v_i. Hence it is obvious how to utilize J_i for $i \in s$ in unbiasedly estimating μ and the variance of the estimator of μ.

Takahasi and Sakasegawa's Scheme Modified by Pal

Takahasi and Sakasegawa's (1977) scheme involves an implicit randomization rather than an RR technique. Here also the problem is to unbiasedly estimate θ, the proportion bearing A. A sampled person is to make a silent choice of the top priority of oneself among the colors, say, violet, blue, and green. Presuming that everyone has such a definite choice is to respond 1 or 0 as per the display in Table 3.1 depending on the person's bearing A/A^c and top choice favoring one of the three colors. For the estimation of θ, three independent samples are to be chosen as say, s_1, s_2, s_3. A person sampled is to respond on implementing the scheme, as in the Table 3.1.

Alternative schemes are possible, but this admits an unbiased estimator for θ in the following way:

TABLE 3.1

Guided Reporting Sample-Wise

Preferred Color	s_1 A/A^c		s_2 A/A^c		s_3 A/A^c	
Violet (V)	0	1	1	0	1	0
Blue (B)	1	0	0	1	1	0
Green (G)	1	0	1	0	0	1

Let θ_{AV}, θ_{AB}, θ_{AG} be proportions bearing A with top priority for V, and for B, and G, respectively. Replacing A by A^c we denote the corresponding proportions for those bearing A^c. Let

$$\theta = \pi_{AV} + \pi_{AB} + \pi_{AG}.$$

Let $\lambda_i = Prob[$Response is 1 for one in $s_i]$.

Then,

$$\lambda_1 = \pi_{A^cV} + \pi_{AB} + \pi_{AG}, \quad \lambda_2 = \pi_{AV} + \pi_{A^cB} + \pi_{AG}, \quad \lambda_3 = \pi_{AV} + \pi_{AB} + \pi_{A^cG},$$

giving $\lambda_1 + \lambda_2 + \lambda_3 = \theta + 1$.

Writing $\hat{\lambda}_i$, $i = 1, 2, 3$, the corresponding sample proportions, an unbiased estimator for θ follows as $\hat{\theta} = \Sigma_{i=1}^{3} \hat{\lambda}_i - 1$, provided one takes an SRSWR in say, n draws in each sample. Then

$$V(\hat{\theta}) = \frac{1}{n} \sum_1^3 \lambda_i(1-\lambda_i) \quad \text{and} \quad v = \frac{1}{(n-1)} \sum_1^3 \hat{\lambda}_i\left(1 - \hat{\lambda}_i\right),$$

unbiasedly estimates $v(\hat{\theta})$.

Pal (2007a) extends to general sampling schemes on denoting $y_{iv} = 1$ if i bears A and likes violet most, so that $\pi_{AV} = 1/N \Sigma_1^N y_{iv}$. Similarly, y_{iB}, y_{iG} are defined with $\pi_{AB} = 1/N \Sigma_1^N y_{iB}$, $\pi_{AG} = 1/N \Sigma_1^N y_{iG}$. Also, $y_i = y_{iv} + y_{iB} + y_{iG}$ because for anyone only one of the three colours is the top most choice. Of course, $\theta = \pi_{AV} + \pi_{AB} + \pi_{AG}$ and $\pi_{A^cV} = (1 - \pi_{AV})$, $\pi_{A^cB} = (1 - \pi_{AB})$, and so on.

Instead of taking three independent samples, let any person labeled i on being chosen by a general sampling scheme make one of the choices corresponding to $s_1, s_2,$ or s_3 as in Table 3.1 rather with probabilities $p_1, p_2,$ or p_3 such that $p_1 + p_2 + p_3 = 1$.

Let $r'_{iV} = 1$ if ith person with violet as the top priority report 1/0 using
    Table 3.1
    $= 0$, else.

Similarly, r'_{iB}, r'_{iG} are defined. Let

$$r'_{1i} = r'_{1iV} + r'_{1iB} + r'_{1iG}.$$

This exercise is repeated independently for the same person labeled i. Then let $r''_{1i} = r''_{1iV} + r''_{1iB} + r''_{1iG}$ analogous to r'_{1i}.
    Let

$$r_{1i} = \frac{1}{2}(r'_{1i} + r''_{1i}), \quad v_R(r_{1i}) = \frac{(r'_{1i} - r''_{1i})^2}{4}.$$

Then,

$$E_R\left(r_{1i}\right) = E_R\left(r'_{1i}\right) = E_R\left(r''_{1i}\right)$$
$$= p_1\left(1 - y_{iV} + y_{iB} + y_{iG}\right) + p_2\left(y_{iV} + 1 - y_{iB} + y_{iG}\right)$$
$$+ p_3\left(y_{iV} + y_{iB} + 1 - y_{iG}\right) = M_{1i},$$

say, $i \in U$.

$$E_R V_R(r_{1i}) = V_R(r_{1i}) = V_i.$$

Now using r_{1i}, $V_R(r_{1i})$, the estimators e and v follow as usual.
   Then,

$$E(e) = p_1\left(\pi_{A^cV} + \pi_{AB} + \pi_{AG}\right) + p_2\left(\pi_{AV} + \pi_{A^cB} + \pi_{AG}\right) + p_3\left(\pi_{AV} + \pi_{AB} + \pi_{A^cG}\right).$$

$V(e)$ follows as usual. Noting that e is based on the first sample s_1 drawn by a design p, let a second sample s_2 be independently drawn by the same design p. Let, similar to r'_{1i}, r'_{2i} be the two repetitions of the reporting for Ii in s_1; a person Ii, if sampled also in s_2 yields independent reports r''_{1i}, r''_{2i} but using the scheme of reporting as in Table 3.2.

$$p_1 + p_2 + p_3 = 1.$$

Let

$$r_{2i} = \frac{r'_{2i} + r''_{2i}}{2}, \quad v_R\left(r_{2i}\right) = \frac{1}{4}\left(r'_{2i} - r''_{2i}\right)^2.$$

Then

$$E_R\left(r_{2i}\right) = p_1\left(y_{iV} + y_{iB} + 1 - y_{iG}\right) + p_2\left(1 - y_{iV} + y_{iB} + y_{iG}\right)$$
$$+ p_3\left(y_{iV} + 1 - y_{iB} + y_{iG}\right)$$
$$= E_R\left(r'_{2i}\right) = E_R\left(r''_{2i}\right), \quad E_R V_R\left(r_{2i}\right) = V_R\left(r_{2i}\right) i \in U.$$

TABLE 3.2

Reporting by a Person in s_2 Chosen According to Design p

	I A/A^c		II A/A^c		III A/A^c	
Violet (V)	1	0	0	1	1	0
Blue (B)	1	0	1	0	0	1
Green (G)	0	1	1	0	1	0
Selection probability	p_1		p_2		p_3	

TABLE 3.3

Guided Reporting by One in s_3

	I A/A^c		II A/A^c		III A/A^c	
Violet (V)	1	0	1	0	0	1
Blue (B)	0	1	1	0	1	0
Green (G)	1	0	0	1	1	0

Denoting e and v based on s_1 by e_1, v_1 let those based on p_2 be denoted by e_{21}, v_2. Then

$$E(e_2) = p_1(\pi_{AV} + \pi_{AB} + \pi_{A^cG}) + p_2(\pi_{A^cV} + \pi_{AB} + \pi_{AG}) + p_3(\pi_{AV} + \pi_{A^cB} + \pi_{AG}).$$

Let a third sample s_3 be drawn independently, again employing the same sampling design p and a person sampled be requested to report according to the scheme as in the Table 3.3, choosing one of I, II, and III with probability $p_1, p_2,$ or p_3, respectively, such that $p_1 + p_2 + p_3 = 1$.

As in the two earlier cases they are now independently generated r'_{3i}, r''_{3i}, and $r_{3i} = 1/2(r'_{3i} + r''_{3i})$, $v_R(r_{3i}) = 1/4(r'_{3i} - r''_{3i})^2$.

Then,

$$E_R(r_{3i}) = p_1\left(y_{iV} + y_{iB} + 1 - y_{iG}\right) + p_2\left(y_{iV} + y_{iB} + 1 - y_{iG}\right)$$
$$+ p_3\left(1 - y_{iV} + y_{iB} + y_{iG}\right),$$

and $E_R v_R(r_{3i}) = V_R(r_{3i})$.

So e_3, v_3 are now calculated in the same way as e_{31}, v_1 and e_2, v_2.

Then,

$$E(e_3) = p_1\left(\pi_{AV} + \pi_{A^cB} + \pi_{AG}\right) + p_2\left(\pi_{A^cV} + \pi_{AB} + \pi_{A^cG}\right) + p_3\left(\pi_{A^cV} + \pi_{AB} + \pi_{AG}\right).$$

Finally, $e = e_1 + e_2 + e_3$, such that $E(e) = E(e_1) + E(e_2) + E(e_3) = 3 + \theta$.

So finally, $\hat{\theta} = e - 3$ is Pal's (2007a) proposed unbiased estimator for θ. Then, $v = (v_1 + v_2 + v_3)$ gives an unbiased estimator for $V(\hat{\theta})$.

Liu, Chow, and Mosley's RRT

Liu et al. (1975) considered a population divided into T disjoint proportions $\pi_1, \ldots, \pi_j, \ldots, \pi_T (j = 1, \ldots, T)$ with $\sum_1^T \pi_j = 1, 0 < \pi_j < 1, j = 1, \ldots, T$, giving T nonoverlapping groups.

A person is inclined to hide from others if belonging to some of these T groups. Hence, it is advisable to apply RRT by taking $m_1 \neq \cdots \neq m_j \neq \cdots \neq m_T$

beads of T different colors, each representing one of these groups: $m = \Sigma_{0=1}^{T} m_j$ beads are then put into a transparent flask with a long neck and a cork stopper. When the flask is full with beads, it is closed by using the stopper and the flask is turned upside down, so that the beads do not fall out but instead rest on its adjacent bead or on the stopper. The neck of the flask which is marked from $1, 2, \dots, k$ enables one to observe the number against the bead of each color. The person to be sampled is requested to shake the flask, turn it upside down, and give the RR as the number on the neck, which corresponds to the bead of the person's group's color, which is the bottom-most among all the beads of the person's group.

Chaudhuri (2002) gives the following:

$$p_{11} = Prob[\text{a respondent of category 1 reports 1}] = \frac{m_1}{m}$$

$$p_{23} = Prob[\text{a respondent of category 2 reports 3}]$$

$$= \left(\frac{m - m_2}{m}\right)\left(\frac{m - m_2 - 1}{m - 1}\right)\frac{m_2}{m - 2} \quad \text{and so on.}$$

$$p_{jk} = Prob[\text{a respondent of category } j \text{ reports } k], \text{ calculated likewise.}$$

Then, $\lambda_j = Prob[\text{a respondent reports } j] = \sum_{k=1}^{T} p_{kj} \pi_k$

$$j = 1, \quad M = \max_{1 < k < T}(m - m_k + 1).$$

Liu et al. (1975) take an SRSWR in n draws and unbiasedly estimates by λ_j = the observed sample proportion reporting j.

Chaudhuri and Mukerjee (1988) have illustrated how by inverting

$$\hat{\lambda}_j = \sum_{k=1}^{T} p_{kj} \hat{\pi}_k,$$

estimates $\hat{\pi}_k$ for π_k may be worked out by the method of moments. Chaudhuri (2002) extends it to the general scheme of sampling introducing the notations given below:

$$I_{ij} = 1 \quad \text{if } i \text{ reports } j$$
$$= 0 \quad \text{if } i \text{ reports other than } j.$$

Then, $E_R(I_{ij}) = \lambda_j \quad \forall \quad i \in U.$

Chaudhuri (2002) recommends requesting every sampled person Ii for an independently produced second response j, to define

$$I'_{ij} = \text{similar to } I_{ij} \text{ for the second trial.}$$

Then

$$J_{ij} = \frac{1}{2}(I_{ij} + I'_{ij}) \text{ has } E_R(J_{ij}) = \lambda \; \forall i,$$

and

$$V_{ij} = V_R\left(J_{ij}\right) = \frac{V_R\left(I_{ij}\right)}{2},$$

and

$$v_{ij} = \frac{1}{4}\left(I_{ij} - I'_{ij}\right)^2 \forall i,$$

an unbiased estimator for V_{ij}.

Then, following Chaudhuri and Mukerjee, $\lambda_j = \sum_{k=1}^{T} p_{kj}\pi_k$ may be estimated unbiasedly by $J_{ij} \forall i$ and also by $e_j = 1/N \sum_i J_{ij}b_{si}$ calculated as for e earlier. Similarly, like v, we may calculate v_j by the same approach starting with J_{ij} for $i \in s$.

4

Maximum Likelihood Approach

Introduction

Let us recall our definitions:

$$y_i = 1 \quad \text{if } i\text{th person bears the attribute A}$$
$$= 0, \quad \text{else, } i \in U, \quad U = (1,\ldots, i,\ldots, N),$$

r_i is an unbiased estimator for y_i, that is, $E_R(r_i) = y_i$ with a variance $V_R(r_i) = V_i$, say, $i \in U$. For $Y = \sum_1^N y_i$ an unbiased estimator in terms of r_i is derived on starting with $t = \sum_{i \in s} y_i b_{si}$, such that b_{si} is free of $\underline{Y} = (y_1, \ldots, y_i, \ldots, y_N)$ and $\underline{R} = (r_1, \ldots, \overline{r}_i, \ldots, r_N)$ and $\sum_{s \ni i} b_{si} \, p(s) = 1 \; \forall i$. Then,

$$\hat{t} = \sum_{i \in s} r_i b_{si} \tag{4.1}$$

is an unbiased estimator for Y because $E(\hat{t}) = E_p E_R(\hat{t}) = E_R E_p(\hat{t}) = Y$.

Writing

$$V_p(t) = \sum y_i^2 C_i + \sum_{i \neq} \sum_j y_i y_j C_{ij},$$

with

$$C_i = \sum_{s \ni i} b_{si}^2 p(s) - 1,$$

and

$$C_{ij} = \sum_{s \ni i, j} b_{si} \, b_{sj} p(s) - 1,$$

one has

$$V(\hat{t}) = V(t) + \sum V_i \left[\sum_{s \ni i} b_{si}^2 p(s) \right]$$
$$= V(t) + \sum V_i (1 + C_i).$$

Writing

$$v_p(t) = \sum_{i \in s} y_i^2 C_{si} + \sum_{i \neq j}\sum_{\in s} y_i y_j C_{sij},$$

such that

$$\sum p(s) C_{si} = C_i,$$

and

$$\sum_s p(s) C_{sij} = C_{ij},$$

it follows that

$$v_1(\hat{t}) = v_p(t)\Big|_{\underline{Y}=\underline{R}} + \sum_{i \in s}\hat{V}_i b_{si} \qquad (4.2)$$

and

$$v_2(\hat{t}) = v_p(t)\Big|_{\underline{Y}=\underline{R}} + \sum_{i \in s}\hat{V}_i\left(b_{si}^2 - C_{si}\right) \qquad (4.3)$$

are two unbiased estimators for $V(\hat{t})$, provided $\hat{V}_i, b_{si}, C_{si}, C_{sij}$ may all be found out with the desired properties noted above and in addition $E_R(\hat{V}_i) = V_i$ also is satisfied.

Next, our plan is to employ alternatives to $\hat{t}, v_1(\hat{t}), v_2(\hat{t})$ on replacing r_i by maximum likelihood estimators (MLE) of y_i derived suitably for respective randomized response (RR) devices as we shall illustrate next.

Illustrations

Warner's Model

As our mission in this book is to succinctly propagate the central message that no randomized response technique (RRT) needs to be tied to any particular way a sample of persons is to be addressed to extract a response; for several illustrated RRT's we show how MLEs for y_i for any i may be derived to produce estimators of the forms 4.1 through 4.3.

From Warner's (1965) RRT, we may recall that

$I_i = 1$ if card type "matches" the ith person's true feature A or A^c
$\quad = 0$, if it "does not match."

Then,

$$Prob(I_i = 1) = py_i + (1 - p)(1 - y_i)$$
$$= (1 - p) + (2p - 1)y_i = E_R(I_i).$$

Also, we have the "conditional probabilities"

$$Prob(I_i = 1 \mid y_i = 1) = p, \quad Prob(I_i = 1 \mid y_i = 0) = 1 - p,$$
$$Prob(I_i = 0 \mid y_i = 1) = 1 - p, \quad Prob(I_i = 0 \mid y_i = 0) = p.$$

Also, $E_R(1 - I_i) = p - (2p - 1)y_i$. So, $Prob(I_i = y_i) = p$, $Prob(I_i = 1 - y_i) = 1 - p$. So, we may write down the likelihood of y_i given I_i as

$$L(y_i \mid I_i) = \left[p^{y_i}(1 - p)^{1 - y_i} \right]^{I_i} \left[p^{1 - y_i}(1 - p)^{y_i} \right]^{1 - I_i}$$
$$= p^{I_i y_i + (1 - I_i)(1 - y_i)}(1 - p)^{I_i(1 - y_i) + y_i(1 - I_i)}$$

Then, $L(I_i \mid I_i) = p$, $L(1 - I_i \mid I_i) = (1 - p)$, because $y_i^2 = y_i$ and $I_i^2 = I_i$.
So, the MLE of y_i based on Warner's (1965) RRT-based data is

$$m_i = I_i \qquad \text{if } p > \frac{1}{2}$$
$$= 1 - I_i \quad \text{if } p < \frac{1}{2}$$

Note: If $p = 1/2$, no MLE for y_i exists.

Let $I_{si} = 1/0$ if $s \ni i/s \not\ni i$ for a sample s chosen with a probability $p(s)$ such that $\sum_{s \ni i} p(s)b_{si} = 1 \forall i$, we may take $e = \sum m_i b_{si} I_{si}$ to claim $E_p(e) = \sum_1^N m_i$.
Since

$$E_R(m_i) = (1 - p) + (2p - 1)y_i, \quad \text{if } p > 1/2$$
$$= p - (2p - 1)y_i, \quad \text{if } p < \frac{1}{2}$$

one gets

$$E(e) = N(1 - p) + (2p - 1)Y$$
$$= N\left[(1 - p) + (2p - 1)\theta \right], \quad \text{if } p > \frac{1}{2},$$

or

$$E(e) = N\left[p - (2p - 1)\theta\right], \quad \text{if } p < \frac{1}{2}.$$

Therefore,

$$f = \frac{1}{(2p - 1)}\left[\frac{e}{N} - (1 - p)\right], \quad \text{if } p > \frac{1}{2}$$

is an unbiased estimator for θ or

$$g = \frac{1}{(2p - 1)}\left(p - \frac{e}{N}\right),$$

if $p < 1/2$, is an unbiased estimator for θ. So, if guided by a predilection for the "Likelihood Approach" one may propose f if $p > 1/2$, or g if $p < 1/2$ as an alternative unbiased estimator for θ to compete with $\hat{t} = \sum_1^N r_i b_{si} I_{si}$ given by Warner.

For m_i which satisfies $m_i^2 = m_i$ we may note $V_R(m_i) = E_R(m_i)(1 - E_R(m_i)) = p(1 - p) = V_i$, say, for $i \in U$.

Further, $E_R(e) = \sum E_R(m_i) b_{si} I_{si}$, $V_R(e) = \sum V_i b_{si}^2 I_{si}$. Therefore,

$$V(e) = \sum E_R^2(m_i) C_i + \sum_{i \neq} \sum_j E_R(m_i) E_R(m_j) C_{ij} + \sum V_i C_i + \sum V_i$$

and also

$$V(e) = E_p \sum V_i b_{si}^2 I_{si} + V_p\left(\sum E_R(m_i) b_{si} I_{si}\right).$$

Letting

$$v_p(e) = \sum_{i \in s} m_i C_{si} + \sum_{i \neq} \sum_{j \in s} m_i m_j C_{sij},$$

it follows that $v_1 = v_p(e) + \sum V_i b_{si} I_{si}$ and $v_2 = v_p(e) + \sum V_i(b_{si}^2 - C_{si}) I_{sij}$ follow as two unbiased estimators for $V(e)$.

Since

$$V(f) = \frac{V(e)}{N^2 (2p - 1)^2} = V(g),$$

for g and f two unbiased estimators follow as

$$v_1(f) = \frac{v_1}{N^2(2p-1)^2} = v_1(g)$$

and

$$v_2(f) = \frac{v_2}{N^2(2p-1)^2} = v_2(g)$$

Note: It may be remarked that the literature on survey sampling abounds with examples of $p(s)$, b_{si}, C_{si}, C_{sij} and we only need to refer to Chaudhuri and Stenger (2005) as a ready reference.

Mangat and Singh's Model

As mentioned in Chapter 3, $T(0 < T < 1)$ is the probability that every person on request gives out one's true A/A^c characteristics; on drawing from a given box with cards marked T denoting an instruction for true feature and the secondbox with cards marked R with an instruction to report on following Warner's RR device. The proportions of cards marked T or R in the first box are $T(0 < T < 1)$ and $(1 - T)$, respectively. Everything else is in accordance with Warner's RRT. Then, it easily follows that

$$Prob(I_i = y_i) = T + (1-T)p = 1 - \alpha$$

say,

$$\begin{aligned} E_R(I_i) &= \alpha + (1-2\alpha)y_i \\ &= (1-T)(1-p) + \left[T + (1-T)(2p-1)\right]y_i \end{aligned}$$

leading to

$$r_i = \frac{I_i - (1-T)(1-p)}{T + (1-T)(2p-1)}, \quad E_R(r_i) = y_i$$

The likelihood of y_i given I_i is

$$L(y_i \mid I_i) = \left[(1-\alpha)^{y_i}\,\alpha^{(1-y_i)}\right]^{I_i}\left[\alpha^{y_i}(1-\alpha)^{(1-y_i)}\right]^{1-I_i}$$

So, the MLE of y_i given I_i is

$$m_i = I_i \qquad \text{if } \alpha < \frac{1}{2}$$
$$= 1 - I_i \quad \text{if } \alpha > \frac{1}{2}$$

No MLE exists if $\alpha = 1/2$.
  Also,

$$V_i = V_R(m_i) = E_R(m_i)\big(1 - E_R(m_i)\big)$$
$$= (1-T)(1-p)\big[1 - (1-T)(1-p)\big] \quad \forall i \in U.$$

Therefore, proceeding as in Warner's (1965) model an unbiased estimator for θ and two unbiased estimators for the variance of that estimator in terms of m_i, b_{si}, C_{si}, C_{sij} follow rather quite apparently.

Kuk's Model

Recalling Kuk's RR model covered in Chapter 3 giving f_i as the number of "red" cards observed in K draws from a box containing "red" and "nonred" cards in proportions $\theta_1 : \theta_2, (0 < \theta_1 < 1; 0 < \theta_2 < 1)$, the likelihood of y_i, given f_i from a respondent labeled i, is $L(y_i \mid f_i) = [\theta_1^{f_i}(1-\theta_1)^{K-f_i}]^{y_i}[\theta_2^{f_i}(1-\theta_2)^{K-f_i}]^{1-y_i}$.
  Then, if $\theta_1 = \theta_2$, no MLE exists for y_i; otherwise, the MLE of y_i is

$$m_i = 1 \quad \text{if } \left(\frac{\theta_1}{\theta_2}\right)^{f_i} > \left(\frac{1-\theta_1}{1-\theta_2}\right)^{K-f_i},$$

that is, if

$$f_i > K\,\frac{\log\dfrac{(1-\theta_1)}{(1-\theta_2)}}{\log\dfrac{[\theta_1(1-\theta_1)]}{[\theta_2(1-\theta_2)]}}$$

and $m_i = 0$ when the above inequality is reversed.
  Since we are unable to offer in simple analytic forms any formulae for $E_R(m_i)$ and $V_R(m_i)$ we cannot present unbiased estimator for θ and corresponding unbiased variance estimators, our proposal is the following.
  By independently implementing the Kuk's RRT once again for respondent i in the first sample by the same method a second MLE, say, m_i' be derived for y_i.

Now, let

$$\psi_i = \frac{1}{2}(m_i + m_i');$$

then

$$V_R\psi_i = \frac{V_R(m_i)}{2} = V_i,$$

say, has an unbiased estimator

$$v_i = \frac{1}{4}(m_i - m_i')^2,$$

for itself.
   Let

$$e = \frac{1}{N}\sum \psi_i b_{si} I_{si}$$

be a proposed estimator for θ, treating $E_R(\psi_i)$ as an approximation for y_i.
   Finally, we propose

$$v_1 = \frac{1}{N^2}\left[\sum \psi_i^2 C_{si} I_{si} + \sum_{i\neq}\sum_j \psi_i\psi_j C_{sij} + \sum v_i b_{si} I_{si}\right]$$

and

$$v_2 = \frac{1}{N^2}\left[\sum \psi_i^2 C_{si} I_{si} + \sum_{i\neq}\sum_j \psi_i\psi_j C_{sij} + \sum v_i (b_{si}^2 - C_{si}) I_{si}\right]$$

as two approximately unbiased estimators for $V(e)$.

Unrelated Characteristics Model

In Chapter 3, we described the unrelated characteristics model and the RRTs concerned, as introduced by Horvitz et al. (1967) and Greenberg et al. (1969) related to SRSWR and further researched by Chaudhuri (2001a) to apply to general sampling designs. We show here how MLEs may be derived in relation to these RRTs.
   Let us recall that y_i is as before valued 1/0 as the ith person bearing the sensitive attribute A or its complement A^c, respectively. Also, x_i is 1/0-valued

if i bears an innocuous and unrelated feature B or the latter's complement B^c. A box with similar cards marked A or B in proportions $p_1:(1-p_1)$, $(0 < p_1 < 1)$ and a second box marked likewise but in proportions $p_2:(1-p_2)$, $(0 < p_2 < 1, p \neq p_2)$ are to be used independently to produce by i an RR as I_i, J_i for the first and the second box, respectively, with 1/0 values each for a match/no match of the card type versus the genuine A/B attributes.

Then, the joint likelihood of (y_i, x_i) given I_i is

$$L\left(y_i, x_i \mid I_i\right) = \left[p_1 y_i + \left(1-p_1\right)x_i\right]^{I_i}\left[p_1\left(1-y_i\right)+\left(1-p_1\right)\left(1-x_i\right)\right]^{1-I_i}.$$

Then,

$$L\left(I_i, I_i \mid I_i\right) = I_i^{I_i}\left(1-I_i\right)^{1-I_i}$$
$$= 0 \quad \text{if } I_i = 1,$$
$$= 0, \quad \text{if } I_i = 0;$$

$$L(I_i, 1 - I_i \mid I_i) = p_1 \quad \text{both for} \quad I_i = 1, 0;$$

$$L\left(1-I_i, I_i \mid I_i\right) = 1-p_1 \quad \text{if } I_i = 1$$
$$= 1-p_1 \quad \text{also if } I_i = 0;$$

$$L\left(1-I_i, 1-I_i \mid I_i\right) = 0 \quad \text{both for } I_i = 1,0.$$

Then, the MLE of y_i, namely m_i is $m_i = I_i$ for every i in a sample s, given the RR as I_i.

Similarly, given J_i, the MLE of y_i is g_i, which equals J_i. So, it may be easily checked that

$$r_i = \frac{\left(1-p_2\right)m_i - \left(1-p_1\right)g_i}{\left(p_1 - p_2\right)}$$

has $E_R(r_i) = y_i$, because

$$E_R\left(m_i\right) = E_R\left(I_i\right) = p_1 y_i + \left(1-p_1\right)x_i \text{ and } E_R\left(g_i\right) = E_R\left(J_i\right) = p_2 y_i + \left(1-p_2\right)x_i.$$

Now

$$V_i = V_R\left(r_i\right) = E_R\left(r_i^2\right) - E_R\left(r_i\right)$$
$$= E_R\left[r_i\left(r_i - 1\right)\right],$$

because $y_i^2 = y_i$.

So, $\widehat{V}_i = r_i(r_i - 1)$ satisfies $E_R(\widehat{V}_i) = V_i$.

So, $e = 1/N \sum r_i b_{si} I_{si}$ is our proposed unbiased estimator for θ based on the MLE's m_i and g_i for y_i, $i \in s$. Two unbiased estimators for $V(e)$ are then

$$v_1 = \frac{1}{N^2}\left[\sum r_i^2 C_{si} I_{si} + \sum_{i \neq} \sum_j r_i r_j C_{sij} I_{sij} + \sum \widehat{V}_i b_{si} I_{si}\right]$$

and

$$v_2 = \frac{1}{N^2}\left[\sum r_i^2 C_{si} I_{si} + \sum_{i \neq} \sum_j r_i r_j C_{sij} I_{sij} + \sum \widehat{V}_i \left(b_{si}^2 - C_{si}\right) I_{si}\right]$$

We omit the similar exercises that may be pursued with the models given by Mangat (1992) and Mangat et al. (1992) tied to SRSWR and later released by Chaudhuri (2001a) to general sampling designs.

5

Optional RRT

Introduction

In the actual application of the randomized response (RR) techniques known till the early 1980s, in a small survey of the postgraduate students in various education centers in Kolkata, under my guidance/but implemented by the trainees as fieldworkers of the Indian Statistical Service (ISS), quite a few male and female students gave direct responses (DRs) about their habits anticipated to bear social stigmas, ignoring the RR devices offered to them to ensure secrecy.

This led to the first publication of an optional randomized response (ORR) technique by Chaudhuri and Mukerjee (1985). In their book Chaudhuri and Mukerjee (1988) gave more details, both linked to SRSWR. So, we omit their repetition here. Chaudhuri and Saha (2005) released it to the wide class of general sampling designs. In all these cases, the respondents directly divulged their truths if they chose to do so, while others applied RR devices.

Chronologically, to our knowledge, Mangat and Singh (1994), Singh and Joarder (1997), Gupta et al. (2002), Arnab (2004), and Pal (2008) dealt with the alternative ORR techniques of permitting secrecy about the DR or the RR option actually exercised. The first three of these articles were restricted to SRSWR plus the assumption of the existence of an unknown proportion of people in the community seemed to be willing to give out the direct truth. Arnab (2004) claimed that each person should have a separate probability unknown to the investigator to go for the direct revelation of the personal feature. But his detailed write-ups applied to unequal probability sampling really follow Chaudhuri and Saha's (2005) model. Pal (2008) covered unequal probability sampling permitting a separate unknown probability for each respondent to go for a DR rather than an RR. In a recent article, Chaudhuri and Dihidar (2009) have seemingly presented the ORR technique in its culminating manifestation. In the next section, we illustrate only the developments due to Chaudhuri and Saha (2005) and Chaudhuri and Dihidar (2009) in which the present author actively participated.

Illustrations

Chaudhuri and Saha's ORR Technique

Suppose from a population $U = (1,\ldots, i,\ldots, N)$ a sample s is chosen with a probability $p(s)$. Letting

$$y_i = 1 \quad \text{if } i \text{ bears } A$$
$$= 0 \quad \text{if } i \text{ bears } A^c,$$

our problem is to unbiasedly estimate the total $Y = \sum_1^N y_i$.

If DRs are available as y_is from i in s let an estimator

$$t_b = t_b\left(s, \underline{Y}\right) = \sum y_i b_{si} I_{si}$$

be contemplated for use, writing

$$I_{si} = 1 \quad \text{if } i \in s$$
$$= 0 \quad \text{if } i \notin s$$

and choosing b_{si}'s free of $\underline{Y} = (y_1,\ldots, y_i,\ldots, y_N)$ and of $\underline{R} = (r_1,\ldots, r_i,\ldots, r_N)$ such that with respect to an RR device independent RR are derivable as r_i's such that $E_R(r_i) = y_i$, $V_R(r_i) = V_i$ for which estimators v_i's are available such that $E_R(v_i) = V_i$ or V_i's are known; we shall, for simplicity, write w_i such that

$$w_i = V_i \quad \text{if } V_i \text{ is known}$$
$$= v_i \quad \text{if } V_i \text{ is estimated by } v_i.$$

We further suppose that

$$\sum_s p(s) b_{si} I_{si} = 1 \quad \forall i.$$

Consequently, t_b is unbiased for Y because

$$\sum p(s) t_b\left(s, \underline{Y}\right) = Y \quad \forall \underline{Y}.$$

Suppose (1) y_is are not ascertainable as DR for i in s; instead, only r_i's are available as CRR for i in s. Then,

$$e_b = {}^{t_b}\big|_{\underline{Y}=\underline{R}} = \sum r_i b_{si} I_{si} \text{ satisfies } E_R(e_b) = t_b \text{ and } E_p(e_b) = \sum r_i = R,$$

say and $E(e_b) = E_p E_R(e_b) = Y$ and also, $E(e_b) = E_R E_p(e_b) = Y$ so that e_b is unbiased for Y. Further

$$V(e_b) = E_p V_R(e_b) + V_p E_R(e_b) = E_p\left[\sum_i V_i b_{si}^2 I_{si}\right] + V_p(t_b),$$

with

$$V_p(t_b) = \sum y_i^2 C_i + \sum_{i \neq} \sum_j y_i y_j C_{ij},$$

with

$$C_i = \sum p(s) b_{si}^2 I_{si} - 1$$
$$C_{ij} = \sum p(s) b_{si} b_{sj} I_{sij} - 1.$$

Choosing, in manners discussed in the literature on survey sampling, for example, Chaudhuri and Stenger (2005), C_{si} and C_{sij} such that $\sum p(s) C_{si} = C_i$, $\sum p(s) C_{sij} = C_{ij}$ we get

$$v_p(t_b) = \sum_i y_i^2 C_{si} I_{si} + \sum_{i \neq} \sum_j y_i y_j C_{sij} I_{sij}$$

such that $E_p v_p(t_b) = V_p(t_b)$.

Consequently, an unbiased estimator $\hat{V}(e_b)$ for $V(e_b)$ has two alternative forms

$$v_1(e_b) = v_p(t_b)\Big|_{\underline{Y} = \underline{R}} + \sum w_i b_{si} I_{si}$$

and

$$v_2(e_b) = v_p(t_b)\Big|_{\underline{Y} = \underline{R}} + \sum w_i (b_{si}^2 - C_{si}) I_{si}$$

as two unbiased estimators for $V(e_b)$ as are verifiable, especially following Chaudhuri et al. (2000).

Next, suppose (2) s is found to be composed of two disjoint parts s_1 and s_2 such that ORRs available as y_i for i in s_1 and r_i for i in s_2, then the proposed unbiased estimator for $\underline{Y}$ is

$$e_b^* = \sum_{i \in s_1} y_i b_{si} I_{si} + \sum_{i \in s_2} r_i b_{si} I_{si}$$

This is so, because writing E_{DR} as the conditional expectation operator over the RR device employed, keeping the DRs given as fixed, it follows that

$$E_{DR}(e_b) = e_b^*$$

Then,

$$E_R(e_b^*) = t_b = E_R(e_b).$$

But

$$E_R \left(e_b - e_b^* \right)^2 = E_R \left[\left(e_b - t_b \right) - \left(e_b^* - t_b \right) \right]^2$$
$$= V_R \left(e_b \right) + V_R \left(e_b^* \right) - 2 E_R \left(e_b^* - t_b \right) E_{DR} \left(e_b - t_b \right)$$
$$= V_R \left(e_b \right) + V_R \left(e_b^* \right) - 2 E_R \left(e_b^* - t_b \right)^2$$
$$= V_R \left(e_b \right) - V_R \left(e_b^* \right).$$

So,

$$V_R \left(e_b^* \right) = V_R \left(e_b \right) - E_R \left(e_b - e_b^* \right)^2 \leq V_R \left(e_b \right).$$

So,

$$V \left(e_b^* \right) = E_p V_R \left(e_b^* \right) + V_p E_R \left(e_b^* \right)$$
$$= E_p V_R \left(e_b \right) - E_p E_R \left(e_b - e_b^* \right)^2 + V_p E_R \left(e_b \right)$$
$$= V \left(e_b \right) - E_p E_R \left(e_b - e_b^* \right)^2.$$

Thus, an unbiased estimator for $V(e_b^*)$ is $\hat{V}(e_b^*) = \hat{V}(e_b) - (e_b - e_b^*)^2$.

A further alternative unbiased estimator for $V(e_b^*)$ is also available, because

$$E_R \left(e_b - e_b^* \right)^2 = E_R \left[\sum_{i \in s_1} \left(r_i - y_i \right) b_{si} I_{si} \right]^2$$
$$= \sum_{i \in s_1} V_i b_{si}^2 I_{si}$$

has an unbiased estimator

$$= \sum_{i \in s_1} w_i b_{si}^2 I_{si}.$$

Thus,

$$\hat{V}' \left(e_b^* \right) = \hat{V} \left(e_b \right) - \sum_{i \in s1} w_i b_{si}^2 I_{si},$$

is a resulting alternative unbiased estimator for $V(e_b^*)$.

Chaudhuri and Dihidar's ORR Technique

In this, a respondent is requested either to give the true 1/0 value of the qualitative variable deemed stigmatizing by the researcher or truthfully the value as per the dictate of the RR device employed, without divulging the course between these two alternatives actually followed. Each respondent labeled i ($= 1,\ldots, N$) is deemed to have an unknown probability C_i ($0 \le C_i \le 1$) to opt to give out the true value and ($1 - C_i$) to follow the direction of the RR device used.

Illustrating Warner's (1965) RR device, no matter how a sample is chosen with positive inclusion probability for every person and also for every pair of distinct persons π_i, π_{ij}, respectively, let one box with cards marked A or A^c in proportions $p_1 : (1 - p_1)$, $(0 < p_1 < 1)$ and another similar box with the proportions changed to $p_2 : (1 - p_2)$, $(0 < p_2 < 1, p_1 \ne p_2)$ be offered to each sampled person i. The two boxes must both be used and be independent of each other by every person sampled and addressed. Let the response be I_i on using the first box which is 1/0 accordingly as there is match/mismatch in the card-type and the genuine A/A^c feature. Likewise, let, I_i' be the response on using the second box.

Thus, from the ith person sampled, the ORR is

$$z_i = y_i \text{ with probability } C_i$$
$$= I_i \text{ with probability } (1 - C_i)$$

when the first box is used and the ORR is

$$z_i' = y_i \text{ with probability } C_i$$
$$= I_i' \text{ with probability } (1 - C_i)$$

on using the second box.
Consequently,

$$E_R(z_i) = C_i y_i + (1 - C_i)\left[p_1 y_i + (1 - p_1)(1 - y_i)\right]$$

and

$$E_R(z_i') = C_i y_i + (1 - C_i)\left[p_2 y_i + (1 - p_2)(1 - y_i)\right]$$

leading to

$$E_R\left[(1 - p_2)z_i - (1 - p_1)z_i'\right] = (p_1 - p_2)y_i$$

leading to

$$r_i = \frac{(1 - p_2)z_i - (1 - p_1)z_i'}{(p_1 - p_2)}$$

so that $E_R(r_i) = y_i$, $i \in U$.

Since $y_i^2 = y_i$, $I_i^2 = I_i$, $I_i'^2 = I_i'$, one gets $V_R(r_i) = E_R(r_i^2) - y_i = E_R r_i(r_i - 1)$. So, $r_i(r_i - 1)$ is an unbiased estimator for $V_R(r_i) = V_i$. So, $e_b = \sum r_i b_{si} I_{si}$ unbiasedly estimates $Y = \sum y_i$ and

$$v_1(e_b) = v_p(t_b)\big|_{Y=\underline{R}} + \sum r_i (r_i - 1) b_{si} I_{si}$$

$$v_2(e_b) = v_p(t_b)\big|_{Y=\underline{R}} + \sum r_i (r_i - 1)(b^2_{si} - C_{si}) I_{si}$$

provide unbiased estimators for $V(e_b)$.

Next, let us consider Mangat and Singh's (1990) RR device and the application of ORR in that context.

A sampled person i is given a box with a proportion $T(0 < T < 1)$ of cards marked "True" and the remaining proportion of cards marked "ORR." If a card marked "True" is drawn, the person is to give out the true value y_i and if a card marked "ORR" appears, the person is to respond by exactly following the ORR procedure based on Warner's (1965) basic device with two independent repetitions as described. In no case the box which is used in responding is to be divulged to the enquirer.

Then, the ith person's response is

$$t_i = y_i \text{ with probability } T$$
$$= z_i \text{ with probability } (1 - T)$$

on the basis of the use of the first box with A/A^c-marked cards in proportions of $p_1 : (1 - p_1)$, and $(0 < p_1 < 1)$ if an "ORR" marked card is drawn. Again, if the second box with A/A^c marked cards in proportions of $p_2 : (1 - p_2)$ then $(0 < p_2 < 1, p_1 \neq p_2)$ need to be used in case an "ORR" marked card is drawn in the second trial, the response from the ith person is

$$t_i' = y_i \text{ with probability } T$$
$$= z_i' \text{ with probability } (1 - T)$$

Then, recalling that t_i and t_i' are independent,

$$E_R(t_i) = T y_i + (1 - T)\Big[C_i y_i + (1 - C_i)\big\{p_1 y_i + (1 - p_1)(1 - y_i)\big\}\Big]$$

and

$$E_R(t_i') = T y_i + (1 - T)\Big[C_i y_i + (1 - C_i)\big\{p_2 y_i + (1 - p_2)(1 - y_i)\big\}\Big].$$

Then, $E_R[(1-p_2)t_i - (1-p_1)t_i'] = (p_1-p_2)y_i$. Hence, generically, $r_i = [(1-p_2)t_i - (1-p_1)t']/(p_1-p_2)$ has $E_R(r_i) = y_i$ and $v_i = r_i(r_i-1)$ is an unbiased estimator for $V_R(r_i) = V_i$. Hence, e_b and $v_1(e_b)$, $v_2(e_b)$ emerge as usual.

Next, let us turn to the unrelated question model to see how ORR may apply to it.

Recalling the basic approach of denoting for a sampled respondent i the value $y_i = 1/0$ as i bears A/A^c and $x_i = 1/0$ as i bears B/B^c, where B is an innocuous characteristic unrelated to A, the modified ORR approach is as follows.

Let some person with unknown probability C_i be supposed to truthfully give out the value y_i or with probability $(1-C_i)$ resort to the RR device as follows.

Let the ith person be offered a box with cards marked A/B in proportions $p_1 : (1-p_1)$ and respond I_i as

$$I_i = 1 \quad \text{if card type A/B matches the true feature A/B}$$
$$= 0 \quad \text{if there is no match.}$$

The same sampled person i is to repeat independently the same exercise using the second box with the only change being that the cards marked A/B are now in proportions $p_2 : (1-p_2)$, $(0 < p_1 < 1, 0 < p_2 < 1, p_1 \neq p_2)$, the response now being I_i' instead of I_i of a similar nature on observing a Match/Mismatch.

Thus, the ORR should turn out to be

$$z_i = y_i \quad \text{with probability } C_i, = I_i \text{ with probability } 1 - C_i$$

using the first box and

$$z_i' = y_i \quad \text{with probability } C_i, = I_i' \text{ with probability } 1 - C_i$$

using the outcome of the second box.

Then,

$$E_R(z_i) = C_i y_i + (1-C_i)\left[p_1 y_i + (1-p_1)x_i\right]$$

and

$$E_R(z_i') = C_i y_i + (1-C_i)\left[p_2 y_i + (1-p_2)x_i\right],$$

leading to

$$E_R\left[(1-p_2)z_i - (1-p_1)z_i'\right] = (p_1-p_2)y_i$$

and to

$$r_i = \frac{(1-p_2)z_i - (1-p_1)z_i'}{(p_1-p_2)}, \quad i \in U$$

Then, $E_R(r_i) = y_i$ and $v_i = r_i(r_i - 1)$ which has $E_R(v_i) = V_i = V_R(r_i)$ because $y_i^2 = y_i$. Hence, e_b emerges to unbiasedly estimate $Y = \sum y_i$ and $v_1(e_b), v_2(e_b)$ to unbiasedly estimate $V(e_b)$.

Comments

Any other available RR technique covering a qualitative character deemed stigmatizing as an ORR version should be derivable in similar manners with parallel consequences.

Chaudhuri and Saha (2005) and Chaudhuri and Dihidar (2009) present numerical illustrations with different sampling designs employed in applying some of the ORR techniques they use. So does Pal (2008), whose technique is very similar to that of Chaudhuri and Dihidar (2009). Some are reproduced in Table 10.1. In developing the works relating to all these three papers, the present author did play an active role.

6

Protection of Privacy

Introduction

In their research book, Chaudhuri and Mukerjee (1988) gave an account of how one may calculate the posterior probability of a person given an RR to bear the characteristic A as, say, R which is "Yes" or the characteristic A^c on giving the response as "No." If the "Yes" answer raises the probability of being perceived to bear A, a respondent would naturally dislike giving the "Yes" answer fearing the possibility of disclosure of the sensitive feature consequent on the specific RR and revelation of the secrecy. They covered details available in the relevant works of Lanke (1975b, 1976), Leysieffer (1975), Leysieffer and Warner (1976), and Anderson (1975a,b,c)—all related to SRSWR method of sample selection. Privacy may be protected if a given RR does not enhance the posterior probability beyond the prior probability θ that the person may bear the sensitive attribute A. A thoroughly undesirable aspect of the situation is that "the closer these two probabilities the greater turns out the variance of an unbiased estimator for θ" that is yielded by the specific RR device. After Chaudhuri and Mukerjee's (1988) publication, the only noteworthy work in print on this subject is due to Nayak (1994), which is also related to SRSWR for which on each draw the prior probability of a person's bearing A continues to remain as θ. The first publication in the context covering general probability designs is credited to Chaudhuri and Saha (2004). This is followed by the detailed publication by Chaudhuri, Christofides, and Saha (2009). Some details with illustrations are recorded in this chapter under "Illustrations."

Illustrations

Nayak's (1994) work, briefly recounted in this chapter, succinctly summarizes the earlier works on the subject "Protection of privacy on sensitive issues anticipating revelation through RRs" available till then in the literature. His work is related, as the earlier ones, to SRSWR alone, yet it provides us the real clue to subsequent works permitting sample selection through unequal probabilities even without replacement as a matter of preference.

With obvious notations $a = Prob(Yes/A)$, $b = Prob(No/A^c)$, applying Bayes' theorem and noting prior probabilities $P(A) = \theta$, $P(A^c) = 1 - \theta$, Nayak (1994) notes

$$Prob(A\,|\,Yes) = \frac{a\theta}{a\theta + (1 - b)(1 - \theta)} = P(A\,|\,Yes)$$

$$Prob(A\,|\,No) = \frac{\theta(1 - a)}{\theta(1 - a) + (1 - \theta)b} = P(A\,|\,No)$$

which provide "measures of revelation of secrecy" about a person's true characteristic A or A^c.

If $P(A|R) > \theta$, the response R as Yes/No is "jeopardizing" with respect to A, and if $P(A^c|R) > 1 - \theta$, the response R is "jeopardizing" with respect to A^c. So, he defines a "measure of jeopardy" as

$$J(R) = \frac{P(A\,|\,R)/\theta}{P(A^c\,|\,R)/\,(1 - \theta)}$$

inherent in a response R concerning A or A^c. The higher it is, the more is its deviation from unity, which is undesirable.

Borrowing this idea of Nayak (1994) tied exclusively to the SRSWR situation, Chaudhuri and Saha (2004) and later Chaudhuri et al. (2009) proceed as follows to cover general sampling situations quite elegantly.

Let $L_i(0 < L_i < 1)$ be an unknown prior probability that y_i takes the value "1" for the person labeled i.

Let $L_i(R)$ denote the posterior (i.e., conditional) probability that the person bears the sensitive attribute A rather than A^c when an RR is elicited by persuasion from the person as a response R.

For any specific RR device employed,

$$J_i(R) = \frac{L_i(R)/L_i}{[1 - L_i(R)]/(1 - L_i)}, \quad i \in U$$

is taken as the "jeopardy measure" for the RRT associated with the person i giving out the specific RR as R.

This indicates the risk of revelation of the ith person's status as A emanating from the RR given out as R for the particular RR device accepted by the person as worth cooperating with.

Therefore, before one agrees to apply a specific RR device, this value must be measured by a possible respondent who is thereby to judge the acceptability of the device as one to protect the person's confidentiality despite the RR to the extent rationally possible.

Chaudhuri et al. (2009) rather prefer an average measure

$$\bar{J_i} = \frac{1}{|R|} \sum_R J_i(R)$$

denoted by $\sum_R$, the sum over all the possible RRs envisaged for the specific RR device and by $|R|$, the cardinality of the set of such RRs as possible, which is the total number of possible RRs.

Thus, their proposed "jeopardy measure" $\bar{J_i}$ is "RR device-specific" and not "RR-specific" separately for every respondent to be approached. The closer the value of $\bar{J_i}$ to unity, the more the ith person is to be reassured of the privacy being protected by the implementation of the RR device to be relied upon.

Unfortunately, the RR devices proposed in the literature thus far have a common trait that, the more the privacy is protected, the more the variances of the unbiased estimators for θ display a tendency to go on increasing, a highly discouraging state of affairs, indeed! As given by Chaudhuri et al. (2009), let us reproduce some of the relevant formulae for varying RR devices already narrated in earlier chapters persisting with the notations used therein avoiding repetitive narrations.

1. For Warner's model

$$L_i(1) = \frac{L_i P(I_i = 1 | y_i = 1)}{L_i P(I_i = 1 | y_i = 1) + (1 - L_i) P(I_i = 1 | y_i = 0)}$$

$$= \frac{p L_i}{(1 - p) + (2p - 1) L_i},$$

$$L_i(0) = \frac{L_i P(I_i = 1 | y_i = 1)}{L_i P(I_i = 0 | y_i = 1) + (1 - L_i) P(I_i = 0 | y_i = 0)}$$

$$= \frac{(1 - p) L_i}{p + (1 - 2p) L_i}.$$

As $p \to 1/2$, $L_{i(1)} \to L_i$ and $L_{i(0)} \to L_i$ as are desirable but as a matter of frustration, $V_i \to +\infty$. In addition,

$$J_i(1) = \frac{p}{1 - p}, \quad J_i(0) = \frac{1 - p}{p},$$

$$\bar{J_i} = \frac{1}{2}[J_i(1) + J_i(0)] = \frac{1}{2}\left(\frac{p}{1 - p} + \frac{1 - p}{p}\right).$$

For $p = 1/2$, $J_i(1) = J_i(0) = \bar{J_i} = 1$ as is desirable for an RR device.

2. For Forced Response model

$$L_i\left(1\right) = \frac{\left(1-p\right)L_i}{p_1 + \left(1-p_1-p_2\right)L_i} \rightarrow L_i,$$

as $\left(p_1 + p_2\right) \rightarrow 1$ and so also $V_i \rightarrow +\infty$.

$$J_i\left(1\right) = \frac{1-p_2}{p_1}, \quad J_i\left(0\right) = \frac{p_2}{1-p_1},$$

$$\bar{J}_i = \frac{1}{2}\left[\frac{1-p_2}{p_1} + \frac{p_2}{1-p_1}\right].$$

For $p_1 + p_2 = 1$, $J_i(1) = J_i(0) = \bar{J}_i = 1$

3. For Kuk's RR device

$$L_i\left(f_i\right) = \frac{L_i\left[p_1^{f_i}\left(1-p_1\right)^{k-f_i}\right]}{p_2^{f_i}\left(1-p_2\right)^{k-f_i} + L_i\left[p_1^{f_i}\left(1-p_1\right)^{k-f_i} - p_2^{f_i}\left(1-p_2\right)^{k-f_i}\right]} \rightarrow L_i,$$

as $p_1 \rightarrow p_2$ while in this case $V_i \rightarrow \infty$.
But

$$J_i\left(f_i\right) = \frac{p_1^{f_i}\left(1-p_1\right)^{k-f_i}}{p_2^{f_i}\left(1-p_2\right)^{k-f_i}}.$$

If $p_1 = p_2$, $J_i\left(f_i\right)$ equals "unity."
Next,

$$\bar{J}_i = \frac{1}{\left(k+1\right)}\sum_{f_i=0}^{k} J_i\left(f_i\right).$$

If $p_1 = p_2$, then $\bar{J}_i$ equals "unity."

4. For Unrelated Question model:

Writing RRs as (I_i, k_i), (I_i', k_i') taking the forms (1, 1), (1, 0), (0, 1), and (0, 0) one gets

$$L_i(1,1) = \frac{L_i p_1 p_2}{(1 - p_1)(1 - p_2) + L_i(p_1 + p_2 - 1)}.$$

As $(p_1 + p_2) \to 1$, $L_i(1,1) \to L_i$.

V_i may retain a finite value unlike for the three previous RR devices.

Again,

$$J_i(1,1) = \frac{p_1 p_2}{(1 - p_1)(1 - p_2)}$$

As $p_1 \to p_2$, $J_i(1,1) \to 1$.

Also,

$$\bar{J}_i = \frac{1}{4}\left[J_i(1,1) + J_i(1,0) + J_i(0,1) + J_i(0,0)\right]$$

$$= \frac{1}{4}\left[\frac{p_1 p_2}{(1 - p_1)(1 - p_2)} + \frac{p_1(1 - p_2)}{p_2(1 - p_1)} + \frac{p_2(1 - p_1)}{p_1(1 - p_2)} + \frac{(1 - p_1)(1 - p_2)}{p_1 p_2}\right].$$

As

$$(p_1 + p_2) \to 1, \quad \bar{J}_i \to \frac{1}{2} + \frac{1}{2}\left[\left(\frac{p_1}{1 - p_1}\right)^2 + \left(\frac{1 - p_1}{p_2}\right)^2\right];$$

also in this case $J_i(1,0) \to 1$, $J_i(0,1) \to 1$ but $\bar{J}_i \nrightarrow 1$.

But if $p_1 \to p_2$, then $\bar{J}_i \to 1$ but $V_i \to +\infty$ in this case.

5. For Mangat and Singh's (1990) model

$$L_i(1) = \frac{L_i\left[T + (1 - T)p\right]}{(1 - T)(1 - p) + L_i\left[T + (1 - T)(2p - 1)\right]} = \frac{L_i(\phi + \psi)}{\psi + L_i\phi},$$

writing $\psi = (1 - T)(1 - p)$ and $\phi = T + (1 - T)(2p - 1)$ which give $\phi + \psi = T + (1 - T)p$.

For this model

$$V_i = [\psi(1 - \psi)]/\phi^2.$$

So, if $T \to 0$, $p \to 1/2$ and hence $\phi \to 0$, then $L_i(1) \to L_i$, but in this case $V_i \to \infty$.

Also,

$$J_i(1) = 1 + \frac{\phi}{\psi} = 1 + \frac{T + (1 - T)(2p - 1)}{(1 - T)(1 - p)},$$

as $\phi \to 0$, $J_i(1) \to 1$.

Again,

$$\begin{aligned}
\bar{J}_i &= \frac{1}{2}\left[J_i(1) + J_i(0)\right] \\
&= \frac{1}{2}\left[\frac{T + (1 - T)p}{(1 - p)(1 - T)} + \frac{(1 - T)(1 - p)}{T + p(1 - T)}\right] \\
&= \frac{1}{2}\left[\frac{\phi + \psi}{\psi} + \frac{\psi}{\phi + \psi}\right].
\end{aligned}$$

If $\phi \to 0$, $\bar{J}_i \to 1$.

6. For Christofides (2003) model modified by Chaudhuri (2004)

$$\begin{aligned}
L_i(k) &= \frac{L_i P(z_i = k \mid y_i = 1)}{L_i P(z_i = 1 \mid y_i = 1) + (1 - L_i) P(z_i = k \mid y_i = 0)} \\
&= \frac{L_i p_{M+1-k}}{p_k + L_i(p_{M+1-k} - p_k)}
\end{aligned}$$

Also,

$$J_i(k) = \frac{p_{M+1-k}}{p_k}$$

and

$$\bar{J}_i = \frac{1}{M} \sum_{k=1}^{M} \left(\frac{p_{M+1-k}}{p_k}\right).$$

Further,

$$L_i((M + 1)/2) = L_i \quad \text{if M is odd}$$

and

$$J_i((M + 1)/2) = 1 \quad \text{if M is odd.}$$

If

$$p_k \to \frac{1}{M} \quad \forall \ k = 1, \dots, M,$$

then $L_i(k) \to L_i$, $J_i(k) \to 1$, and $\bar{J}_i \to 1$.

But in this case, $V_i \to \infty$.

It is now of interest to examine the numerical values of $L_i(k)$, $J_i(k)$, $\bar{J}_i$, and V_i for various RR devices in competition. This will be explained in Chapter 10 (see, e.g., Tables 10.2 and 10.3).

7

Quantitative Characteristics

Introduction

It may be a good idea to recall a few aspects of the literature concerning the problem of the estimation of $Y = \sum_1^N y_i$ and of the variance or mean square error (MSE) of the estimator of Y when direct response (DR) survey data $(s, y_i | i \in s)$ are available straightaway. Recognizing that y_is in the present context are not directly available but themselves are amenable to unbiased estimation in terms of randomized responses (RR) to be ingeniously procurable by appropriate means that fresh procedures are to be established for the estimation of Y and also of the measures of error of the requisite RR data-based estimators for Y. There is of course a parallelism in the analogous problem in estimation by dint of multistage sampling when y_i is treated as the first-stage unit totals to be estimated through sampling of units in subsequent stages when they are not directly ascertainable. However, there is crucial difference as we shall recount in greater detail in due course. Let us attempt at a review in the next section in quest of a right way to place the reader on the center of the situation in a well-equipped frame, to explore further developments in the right and ever-increasing directions.

Review of Literature

Let $t = t(s, \underline{Y})$ be an estimator for Y based on DR-based survey data $d = (s, y_i | i \in s)$ determined from a sample s chosen from $U = (1, \ldots, i, \ldots, N)$ with probability $p(s)$ by design p such that it involves no y_i for $i \notin s$. By $E_p(t) = \sum p(s)t(s, \underline{Y})$ we mean its design expectation, by $B_p(t) = E_p(t - Y)$ its bias, and by $M_p(t) = E_p(t - Y)^2$ its MSE about Y. Its variance is $V_p(t) = E_p(t - E_p(t))^2$ and $\sigma_p(t) = +\sqrt{V_p(t)}$ its standard error. Clearly, $M_p(t) = \sigma_p^2(t) + B_p^2(t)$. A t is unbiased for Y if $B_p(t) = 0 \forall \underline{Y}$.

Basu (1971) showed that no sampling design worth its name admits an estimator for Y which is unbiased with a variance, which is the smallest possible, uniformly for every $\underline{Y}$. Given a design p, of two unbiased estimators

t_1, t_2 for Y based on it, t_1 is the better if $V(t_1) \leq V(t_2) \forall \underline{Y}$ and $V(t_1) < V(t_2)$ at least for one $\underline{Y}$. Such a t_1 is the best if it is better in this sense than any other unbiased estimator based on p for Y. An unbiased estimator for Y is admissible if no other exists that is better than it. Given the raw survey data $d = (s, y_i | i \in s)$, the reduced data d^* derived from it involving the sample s^* derived from s by ignoring the order and the multiplicity of occurrence of the units in s and taking the y-values of the units in s^* which are the distinct units in s with their order of occurrence suppressed is the "minimal sufficient" statistic corresponding to d.

One consequence of this is that one may construct from $t = t(s, \underline{Y})$ another statistic

$$t^* = \sum_{s \to s^*} p(s) t(s, \underline{Y}) \bigg/ \sum_{s \to s^*} p(s) \tag{7.1}$$

such that $E_p(t) = E_p(t^*)$ and $V_p(t^*) \leq V_p(t)$ and strictly so unless t coincides with t^*; here, by $\sum_{s \to s^*}$ we mean summation over all samples s to which the same s^* corresponds. Within the class of all unbiased estimators a "complete class" is the subclass composed of the unbiased estimators that are "independent of the order and the multiplicity" with which the units occur in the sample on which it is based. This subclass is called a "complete class," because given an estimator outside this subclass there is one inside it, which is better (cf. as t^* above).

Next consider an estimator for Y based on DR-based data $(s, y_i | i \in s)$ as $t_b = \sum_{i \in s} y_i b_{si}$ with b_{si} free of $\underline{Y}$ so that it is a homogeneous linear estimator (HLE) such that it is unbiased in addition, (HLUE) demanding that

$$\sum_{s \ni i} p(s) \, b_{si} = 1 \quad \forall i. \tag{7.2}$$

Suppose s_1, s_2 are two samples such that either (1) $s_1 \cap s_2 = \phi$, the empty set or (2) $s_1 \sim s_2$ that is, a unit of U in one of them is also in the other. A design p that assigns positive selection probability to only such samples as those with the above property is called a unicluster sampling design (UCSD).

Godambe (1955) showed that for a non-UCSD among estimators t_b there does not exist one with the least possible variance for every $\underline{Y}$. For a UCSD, Hanurav (1966), Hege (1965), and Lanke (1975a) showed that among estimators t_b for Y the one having the least variance uniformly in $\underline{Y}$ is

$$t_H = \sum_{i \in s} \frac{y_i}{\pi_i} = \sum_{1}^{N} \frac{y_i}{\pi_i} I_{si}, \tag{7.3}$$

called the Horvitz and Thompson (HT) (1952) estimator; here $\pi_i = \sum_{s \ni i} p(s)$, $I_{si} = 1$ if $i \in s$, $= 0$ if $i \notin s$; incidentally for the existence of an unbiased estimator for Y the design must have $\pi_i > 0 \; \forall i$.

Godambe and Joshi (1965) have shown that for any design the HT estimator is admissible among all unbiased estimators for Y.

In order to effectively discriminate among unbiased estimators for Y and eventually hit upon serviceably optimal ones "superpopulation" models are often postulated treating $\underline{Y} = (y_1,\ldots,y_i,\ldots,y_N)$ as a random vector, and a class of probability distributions' which is referred to as a "Model." By E_m, V_m, C_m we shall mean operators for expectation, variance, and covariance with respect to the probability distribution of $\underline{Y}$ as modeled. We shall presume the operators E_p and E_m as commutative so that the overall expectation and variance operators over the sampling design and the model are

$$E = E_p \, E_m = E_m \, E_p \quad \text{and} \quad V = E_p \, V_m + V_p \, E_m = E_m \, V_p + V_m \, E_p. \qquad (7.4)$$

The same is assumed for E_p and E_R operators as well.

In passing we may also mention that the combination of a design and an estimator based on it, namely $(p.t)$ or (p, t_b) as a "strategy." A strategy (p_1, t_1) is better than another (p_2, t_2), say if $E_{p_1}(t_1) = E_{p_2}(t_2)$, but $V_{p_1}(t_1) \le V_{p_2}(t_2) \; \forall \underline{Y}$ and strictly $V_{p_1}(t_1) < V_{p_2}(t_2)$ at least for one $\underline{Y}$.

Let us illustrate one model for which

$$E_m(y_i) = \mu_i, \quad V_m(y_i) = \sigma_i^2, \quad -\infty < \mu_i < +\infty, \; i \in U, \; \sigma_i > 0 \; \forall i$$

and y_is are independently distributed, $i \in U$. For any design unbiased estimator t for Y, Godambe and Thompson (1977) have shown that

$$E_m V_p(t) \ge \sum \sigma_i^2 \left(\frac{1}{\pi_i} - 1 \right) \ge E_m V_p(t_\mu) \qquad (7.5)$$

here $t_\mu = \sum_{i \in s}(y_i - \mu_i/\pi_i) + \mu$, where $\mu = \sum_1^N \mu_i$.

Basu (1971) called such a t_μ with known $\mu_i s (i \in U)$ a "generalized difference" estimator (GDE). In particular, if W_i's are known real numbers such that $\mu_i = \beta W_i$ with β as an unknown constant, then

$$t_\beta = \sum_{i \in s} \frac{y_i}{\pi_i} + \beta \left(W - \sum_{i \in s} \frac{W_i}{\pi_i} \right), \quad \text{with } W = \sum_{i=1}^N W_i. \qquad (7.6)$$

If, in addition, $\sigma_i = \sigma W_i$ with σ as an unknown positive constant, then one may employ a sampling design allowing a constant number $n(\ge 2)$ of units in every sample s with $p(s) > 0$ such that

$$\pi_i = n \frac{W_i}{W}, \quad i \in U, \text{ then it follows that}$$

$$E_m V_p(t) \ge W\sigma^2 \left(\frac{N-n}{N} \right) = E_m V_{p_{nw}}(t_{HT}) \qquad (7.7)$$

writing $p_{n\omega}$ for a design with $\pi_i = n(W_i/W)$ and p_n a design with n as the fixed size (n) sampling design with n distinct units.

Therefore, under the model specified above, the class of strategies $(p_{n\omega}, t_H)$ is optimal against all the competitors (p_n, t) such that $Ep_n(t) = Y$.

If instead $\sigma_i^2 = \sigma^2 W_i^g$ for $0 \le g \le 2$, then the following interesting situation emerges.

The well-known strategy of Rao, Hartley, and Cochran (RHC) (1962) involves the unbiased estimator $t_{RHC} = \Sigma_n y_i (Q_i/p_i)$ based on the scheme of sampling given by RHC as follows. For the units of U normed size-measures $p_i (0 < p_i < 1, \Sigma_1^N p_i = 1)$ are available. The population is randomly divided into n groups of N_i units, $i = 1, ..., n$, such that $\Sigma_n N_i = N$ writing Σ_n as the sum over the n groups formed. Writing $p_{i1}, ..., p_{ij}, ..., p_{iN_i}$ as the normed size measures of the N_i units falling in the ith group, $Q_i = p_{i1} + \cdots + p_{iN_i}$ and for simplicity y_i, p_i, as the y, p-values for the unit falling in the ith group from which respectively one unit is chosen independently with selection probabilities $p_{ij}/Q_i, j = 1,..., N_i$.

Then,

$$V(t_{RHC}) = A\sum_1^N p_i \left(\frac{y_i}{p_i} - Y\right)^2, \quad A = \frac{\sum_n N_i^2 - N}{N(N-1)} \tag{7.8}$$

Another well-known scheme of sampling by Lahiri, Midzuno, and Sen (LMS) (1951, 1952, 1953) chooses a sample s of n distinct units of U with probability

$$p(s) = \frac{1}{\dbinom{N-1}{n-1}} \sum_{i \in s} p_i = p_{LMS}, \text{ say.} \tag{7.9}$$

The ratio estimator $t_R = \dfrac{\sum_{i \in s} y_i}{\sum_{i \in s} p_i}$ based on this, and p_{LMS} design is unbiased

for Y.   (7.10)

Supposing for HT estimator $\pi_i = nW_iM/W$, $\mu_i = \beta W_i$, $\sigma_i = \sigma W_i^{g/2}$ and for the RHC scheme $N_i = N/n$ for every $i = 1, ..., n$, Chaudhuri and Arnab (1979) have shown that

$$E_m V_{p_{n\omega}} (t_H) \ge E_m V_{p_{LMS}} (t_R) \ge E_m V_{RHC} (t_{RHC}) \quad \text{for } 0 \le g \le 1 \tag{7.11}$$

with the inequalities reversed in the same order for $1 \le g \le 2$ and throughout an equality if $g = 1$.

Since β is unknown, t_β above is not usable in practice. Cassel, Särndal, and Wretman (CSW) (1976) provided a useful alternative, namely

$$t_g = \sum_{i \in s} \frac{y_i}{\pi_i} + \hat{\beta}_Q \left(W - \sum_{i \in s} \frac{W_i}{\pi_i} \right) \tag{7.12}$$

called a generalized regression (Greg) estimator for Y. Here $\hat{\beta}_Q$ is taken as

$$\hat{\beta}_Q = \frac{\displaystyle\sum_{i \in s} y_i W_i Q_i}{\displaystyle\sum_{i \in s} W_i^2 Q_i} \tag{7.13}$$

with Q_i taken as a positive number suitably chosen as one, usually, of

$$Q_i = \frac{1}{W_i}, \ \frac{1}{W_i^2}, \ \frac{1}{\pi_i W_i}, \ \frac{1 - \pi_i}{\pi_i W_i}, \ \frac{1}{W_i^g}, \quad 0 \le g \le 2.$$

The first two and the last one are motivated by $\sigma_i^2 = \sigma^2 W_i$, $\sigma^2 W_i^2$, $\sigma^2 W_i^g$, respectively, because for the illustrated model the least-squares estimator of β is

$$\hat{\beta} = \frac{\displaystyle\sum_{i \in s} y_i W_i / \sigma_i^2}{\displaystyle\sum W_i^2 / \sigma_i^2} \tag{7.14}$$

Hájek (1971) suggested the third choice as it yields

$$\hat{\beta}_Q = \frac{\displaystyle\sum_{i \in s} \frac{y_i}{\pi_i}}{\displaystyle\sum_{i \in s} \frac{W_i}{\pi_i}}$$

and hence

$$t_g = W \frac{\displaystyle\sum_{i \in s} \frac{y_i}{\pi_i}}{\displaystyle\sum_{i \in s} \frac{W_i}{\pi_i}},$$

the ratio estimator for Y. The fourth choice is of Brewer (1979) because of the following reason:

The well-known prediction approach of Brewer (1963) and Royall (1970) among others as further elaborated by Zacks and Bolfarine (1992) predicts Y by

$$\hat{Y}_P = \sum_{i \in s} y_i + \hat{\beta}_Q \left(W - \sum_{i \in s} W_i \right).$$

Both t_g and $\hat{Y}_P$ are generally biased for Y, but t_g is asymptotically design unbiased (ADU) for every Q_i (>0) and $\hat{Y}_P$ is ADU for Y for the fourth choice of Q_i yielding

$$\hat{Y}_B = \sum_{i \in s} y_i + \left(W - \sum_{i \in s} W_i \right) \frac{\sum\limits_{i \in s} y_i (1 - \pi_i)/\pi_i}{\sum\limits_{i \in s} W_i (1 - \pi_i)/\pi_i} = t_B,$$

say, the Brewer's predictor.

Brewer's (1979) ADUness recommends calculating the design expectation of every term in t_g, $\hat{Y}_P$ and equating the result to Y.

Now let us turn to the estimation of variance/MSE. For $t_b = \sum_1^N y_i b_{si} I_{si}$ to estimate Y the MSE is $M = E_p(t_b - y)^2 = \sum_1^N \sum_1^N d_{ij} y_i y_i$, $d_{ij} = E_p(b_{si} I_{si} - 1)(b_{sj} I_{sj} - 1)$. Let it be possible to find constants

$$w_i (\neq 0) \quad \forall i. \tag{7.15}$$

With $z_i = y_i/w_i$, one gets $M = \sum_i \sum_j d_{ij} w_i w_j z_i z_j$.

If $z_i = c \; \forall i \Rightarrow M = 0$ as is true of many t_bs, $M = -\sum_{i<} \sum_j d_{ij} w_i w_j (z_i - z_j)^2$ consequent on $M = -\sum_i \sum_j d_{ij} w_i w_j = 0$ in M as shown by Hájek (1959) and Rao (1979). Consequently, when t_b is unbiased for Y, an unbiased estimator for M, to be written as V, is m, which is to be written as v is

$$-\sum_{i<} \sum_j d_{sij} I_{sij} w_i w_j \left(\frac{y_i}{w_i} - \frac{y_j}{w_j} \right)^2, \tag{7.16}$$

subject to $E_p(d_{sij} I_{sij}) = d_{ij}$; for example, when $d_{sij} = d_{ij}/\pi_{ij}$, provided $\pi_{ij} = \sum_s p(s) I_{sij} > 0 \; \forall i \neq j$ on noting

$$I_{sij} = I_{si} I_{sj} = 1 \quad \text{if both } i, j \in s$$
$$= 0, \text{ otherwise.}$$

If the condition "$z_i = c \forall i \Rightarrow M = 0$" fails, then, writing $\alpha_i = \sum_{j=1}^N d_{ij} w_j$, one gets M as

$$M' = -\sum_{i<} \sum_j d_{ij} w_i w_j \left(\frac{y_i}{w_i} - \frac{y_j}{w_j} \right)^2 + \sum_1^N \frac{y_i^2}{w_i} \alpha_i \tag{7.17}$$

and m as

$$m' = -\sum_{i<}\sum_{j} d_{sij} I_{sij} w_i w_j \left(\frac{y_i}{w_i} - \frac{y_j}{w_j}\right)^2 + \sum \frac{y_i^2}{w_i} \frac{\alpha_i I_{si}}{\pi_i} \tag{7.18}$$

In particular, the HT unbiased estimator $t_H = \sum_{i \in s} y_i/\pi_i$ has the variance

$$V(t_H) = \sum y_i^2 \frac{1 - \pi_i}{\pi_i} + \sum_{i \neq}\sum_{j} y_i y_j \left(\frac{\pi_{ij}}{\pi_i \pi_j} - 1\right)$$

$$= \sum_{i<}\sum_{j} (\pi_i \pi_j - \pi_{ij}) \left(\frac{y_i}{\pi_i} - \frac{y_j}{\pi_j}\right)^2 + \sum \frac{y_i^2}{\pi_i} \beta_i,$$

with

$$\beta_i = \left(1 + \frac{1}{\pi_i} \sum_{j \neq i} \pi_{ij} - \sum \pi_i\right) \tag{7.19}$$

which is zero if every sample has a common number of distinct units $\gamma(s) = \gamma$ when $p(s) > 0$.

As another particular case we cite the RHC, 1962 estimator which has a simple unbiased variance estimator

$$v(t_{RHC}) = \left(\frac{\sum_n N_i^2 - N}{N^2 - \sum_n N_i^2}\right) \sum_n Q_i \left(\frac{y_i}{p_i} - t_{RHC}\right)^2 \tag{7.20}$$

which is uniformly nonnegative.

With these detailed preliminaries, we may turn to "how to deal with quantitative RRs." Much of this has been covered in the Chapter 7 of Chaudhuri and Mukerjee (1988) and in Chapter 12 of Chaudhuri and Stenger (2005).

Two RR devices seem to suit well here.

Device A—Let a person labeled i be offered two boxes, one with cards bearing numbers $a_1, ..., a_L$ with

$$\mu_a = \frac{1}{L} \sum_1^L a_j \neq 0, \quad \sigma_a^2 = \frac{1}{L} \sum_1^L (a_j - \mu_a)^2 > 0$$

and the second with cards bearing numbers $b_1, ..., b_T$ with

$$\mu_b = \frac{1}{T} \sum_1^T b_k, \quad \sigma_b^2 = \frac{1}{T} \sum_1^T (b_k - \mu_b)^2$$

with a request to independently draw one card each from the two boxes, and then return them to the two boxes respectively, after noting the numbers, say a_j from the first and b_k from the second box and report the RR to the investigator, without divulging them, as

$$z_i = a_j y_i + b_k, \quad i \in U \tag{7.21}$$

Consequently,

$$E_R(z_i) = y_i \, \mu_a + \mu_b, \quad V_R(z_i) = y_i^2 \, \sigma_a^2 + \sigma_b^2.$$

Then,

$$r_i = \frac{(z_i - \mu_b)}{\mu_a} \quad \text{has} \quad E_R(r_i) = y_i \tag{7.22}$$

and

$$V_R(r_i) = y_i^2 \left(\frac{\sigma_a^2}{\mu_a^2} \right) + \left(\frac{\sigma_b^2}{\mu_a^2} \right) = V_i, \text{ say.}$$

Then,

$$v_i = \left[r_i^2 \left(\frac{\sigma_a^2}{\mu_a^2} \right) + \frac{\sigma_b^2}{\mu_a^2} \right] \Big/ \left[1 + \frac{\sigma_a^2}{\mu_a^2} \right]$$

$$= \frac{r_i^2 \sigma_a^2 + \sigma_b^2}{\mu_a^2 + \sigma_b^2} \quad \text{has} \quad E_R(v_i) = V_i. \tag{7.23}$$

We shall next write

$$\underline{R} = (r_1, \ldots, r_i, \ldots, r_N) \quad \text{and} \quad R = \sum_1^N r_i \tag{7.24}$$

So, we may derive from the DR-based estimator t_b which is unbiased for Y and is unusable through the RRs, the RR-based estimator $e_b = t_b|_{\underline{Y} = \underline{R}} = \sum_1^N r_i b_{si} I_{si}$ for Y has

$$E(e_b) = E_p \left[E_R(e_b) \right] = E_p(t_b) = Y,$$

and also

$$E(e_b) = E_R \left[E_p(e_b) \right] = E_R(R) = Y.$$

So, we regard e_b as unbiased for Y. Applying analogously for multistage sampling for the RR situation from the approach of Chaudhuri et al. (2000) and Chaudhuri and Pal (2002), we get

$$V(t_b) = E_p \left[V_R(e_b) \right] + V_p \left[E_R(e_b) \right]$$

$$= E_p \left[\sum V_i b_{si}^2 I_{si} \right] + V_p(t_b)$$

$$= \sum V_i \left(1 + C_i\right) - \left[\sum_{i<} \sum_j d_{ij} w_i w_j \left(\frac{y_i}{W_i} - \frac{y_j}{W_j}\right)^2 + \sum \frac{y_i^2}{w_i}\alpha_i\right]$$

writing $C_i = E_p\left(b_{si}^2 I_{si}\right) - 1$.

Then, an unbiased estimator for $M = V(e_b)$ is

$$v_{1r} = v\big|_{\underline{Y}=\underline{R}} + \sum_{i<}\sum_j d_{sij} I_{sij} w_i w_j \left(\frac{v_i}{W_i^2} + \frac{v_j}{W_j^2}\right) - \sum \frac{v_i}{w_i}\alpha_i \frac{I_{si}}{\pi_i} + \sum v_i b_{si}^2 I_{si}$$

Then, $E(v_{1r}) = V(e_b)$.

Alternatively, we get

$$V\left(e_b\right) = E_R\left[V_p\left(e_b\right)\right] + V_R\left[E_p\left(e_b\right)\right]$$

$$= E_R\left[-\sum_{i<}\sum_j d_{ij} w_i w_j \left(\frac{r_i}{w_i} - \frac{r_j}{w_j}\right)^2 + \sum \frac{r_i^2}{w_i}\alpha_i\right] + \sum V_i$$

Now,

$$v_p\left(e_b\right)\big|_{\underline{Y}=\underline{R}} = -\sum_{i<}\sum_j d_{sij} I_{sij} w_i w_j \left(\frac{r_i}{w_i} - \frac{r_j}{w_j}\right)^2 + \sum \frac{r_i^2}{w_i}\alpha_i \frac{I_{si}}{\pi_i}$$

So, for $v_{2r} = v_p(e_b)\big|_{\underline{Y}=\underline{R}} + \sum v_i b_{si} I_{si}$, we get

$$E_p\left(v_{2r}\right) = -\sum_{i<}\sum_j d_{ij} w_i w_j \left(\frac{r_i}{w_i} - \frac{r_j}{w_j}\right)^2 + \sum \frac{r_i^2}{w_i}\alpha_i + \sum_1^N V$$

So, $E(v_{2r}) = V(e_b)$ again.

So, v_{1r} and v_{2r} are two unbiased estimators for $V(e_b)$.

Incidentally, instead of utilizing Hájek's (1959) and Rao's (1979) approach of putting the variance or MSE of an HLUE for Y into the form, which effectively suggests a necessary form of a "uniformly nonnegative unbiased estimator" for itself we may also stick to the original quadratic form

$$M = E_p\left(t_b - Y\right)^2 = \sum_i \sum_j d_{ij} y_i y_j$$

$$= \sum_i C_i y_i^2 + \sum_{i \neq} \sum_j C_{ij} y_i y_j$$

for which an unbiased estimator is

$$m = \sum C_{si} I_{si} y_i^2 + \sum \sum_{i \neq j} C_{sij} I_{sij} y_i y_j \quad \text{with}$$

$$E_p\left(C_{si} I_{si}\right) = C_i \quad \text{and} \quad E_p\left(C_{sij} I_{sij}\right) = C_{ij}.$$

Suppose $E_R(r_i) = y_i$, $V_R(r_i) = V_i$, v_i with $E_R(v_i) = V_i$.

Then for $e_b = \sum r_i b_{si} I_{si}$ corresponding to $t_b = \sum y_i b_{si} I_{si}$ subject to $E_p(b_{si} I_{si}) = 1 \; \forall i$ so that M reduces to the variance of t_b as $V_p(t_b)$ (and m to $v_p(t_b)$) which is the variance of t_b, we have

$$V\left(e_b\right) = V_p\left[E_R\left(e_b\right)\right] + E_p\left[V_R\left(e_b\right)\right] = V_p\left(t_b\right) + E_p\left[\sum V_i b_{si}^2 I_{si}\right] \quad (7.25)$$

and also $V\left(e_b\right) = V_R\left[E_p\left(e_b\right)\right] + E_R\left[V_p\left(e_b\right)\right]$

$$= V_R\left(\sum_1^N r_i\right) + E_R\left[\sum C_i r_i^2 + \sum \sum_{i \neq j} C_{ij} r_i r_j\right]$$

$$= \sum V_i + \left[\left(\sum C_i y_i^2 + \sum \sum_{i \neq j} C_{ij} y_i y_j\right) + \sum C_i V_i\right]$$

$$= \sum V_i \left(1 + C_i\right) + \sum C_i y_i + \sum \sum_{i \neq j} C_{ij} y_i y_j. \quad (7.26)$$

These respectively yield two unbiased variance estimators for e_b as

$$v_3\left(e_b\right) = v_p\left(t_b\right)\big|_{Y=R} + \sum v_i b_{si} I_{si} = v_{3r}$$

$$v_4\left(e_b\right) = v_p\left(t_b\right)\big|_{Y=R} + \sum v_i \left(b_{si}^2 - C_{si}\right) I_{si} = v_{4r}.$$

For the RR device "A" described above, the unbiased estimator r_i has $V_R(r_i) = \alpha y_i^2 + \beta$, $\alpha \neq 0$, $\beta \neq 0$, but there is no term with y_i.

In the RR device "B" described below as introduced by Eriksson (1973, 1980), the independent unbiased estimators r_i for y_i to be derived will have $V_R(r_i)$ as a quadratic form $\alpha y_i^2 + \beta y_i + \psi$, α, β, ψ as known nonzero constant terms. In both A and B, the RRs are independent across the respondents.

Device B—A sampled person labeled i out of $U = (1, \ldots, i, \ldots, N)$ is given a box with a proportion C $(0 < C < 1)$ of cards marked "correct" and more cards bearing numbers $x_1, \ldots, x_j, \ldots, x_T$ in respective proportions $q_1, \ldots, q_j, \ldots, q_T$ such that $\sum_{j=1}^T q_j = 1 - C$, $(0 < q_j < 1, j = 1, \ldots, T)$. The person sampled is requested to draw one of the cards and to report the true value y_i if a "correct" coded card is drawn or the number x_j if a card bearing the number x_j is drawn. The numbers x_j are to be ingeniously chosen so as to cover the anticipated range of the unknown values of y. Then, the RR from i is

$$Z_i = y_i \text{ with probability } C$$
$$= x_j \text{ with probability } q_j.$$

Then,

$$E_R(z_i) = Cy_i + \sum_1^T q_j x_j \quad \text{yielding } r_i = \frac{z_i - \sum_{j=1}^T q_j x_j}{C}, \quad E_R(r_i) = y_i$$

and

$$V_R(r_i) = \frac{1}{C^2} V_R(z_i) = \frac{1}{C^2}\left[Cy_i^2 + \sum_1^T q_j x_j^2 - E_R^2(z_i) \right]$$

$$= \frac{1}{C^2}\left[C(1-C)y_i^2 + \sum q_j x_j^2 - \left(\sum q_j x_j\right)^2 \right]$$

$$\left[\text{writing} - 2C\left(\sum_1^T q_j x_j\right) y_i = \frac{1-C}{C} y_i^2 + \frac{v_q(x)}{C^2} \right.$$

$$\left. \mu_x = \sum q_j x_j / 2q_j \text{ and } v_q(x) = \sum q_j (x_j - \mu_x)^2 / 2q_j \right]$$

$$-\frac{2C\mu_x y_i}{1-C} = \alpha \, y_i^2 + \beta \, y_i + \psi,$$

say, with α, β, ψ as known constants.

For $V_i = \alpha y_i^2 + \beta y_i + \psi$ an obvious unbiased estimator of course is $v_i = (\alpha r_i^2 + \beta r_i + \psi) / 1 + \alpha$, especially observing that $(1 + \alpha) \neq 0$.

A more general RRT of course may produce a more general form of $V_R(r_i)$ as $V_i = \alpha_i y_i^2 + \beta_i y_i + \psi_i$ with α_i, β_i, ψ_i as known constants.

For this RRT as in device "B" estimation of Y and unbiased variance or MSE estimation pose no additional problems as tackled already above.

Before illustrating further specific results, let us turn to the following works in the literature in search of optimal estimators, strategies or nonexistence of the best ones.

First, we concentrate on the works of Adhikary et al. (1984), Chaudhuri (1987), Chaudhuri and Mukerjee (1988), and Sengupta and Kundu (1989).

Starting with $\underline{z} = (z_1, ..., z_i, ..., z_N)$ based on an RRT as described in device "B" above, let $e(s, z)$ be an analog of $e(s, \underline{Y})$ based on DRs such that $E_p[e(s, \underline{Y})] = Y$.

This implies $E_p[e(s,\underline{z})] = \sum_1^N z_i$. Such an $e(s,\underline{z})$ is called a derived estimator corresponding to $e(s,\underline{Y})$. Let us choose an RR-data-based statistic

$$t = t(s,\underline{z}) = \frac{e(s,\underline{z}) - (1-C)\mu_x\left(\sum\limits_{i\in s}\frac{1}{\pi_i}\right)}{C}$$

Then, $E_p(t) = [Z - (1-C)\mu_x N]/C$ and so $E(t) = E_R E_p(t) = Y$, that is, t is pR-unbiased for Y.

We may define t more generally replacing in it $\sum_{i\in s} 1/\pi_i$ by $\sum_{i\in s} b_{si}I_{si}$, both of which are unbiased estimators for N.

For either choice we call the estimator t which is $t(s, z)$ such that it is pR-unbiased for Y. Adhikary et al. (1984) proved Theorem 7.1.

Theorem 7.1:

$E_{pR}h(s,\underline{z}) = E_R E_p h(s,\underline{z}) = 0 \;\forall \underline{Y} \Rightarrow E_p h(s,\underline{z}) = 0 \;\forall\underline{z}$ and   thereby   proved

Theorem 7.2.

Theorem 7.2:

$E_{pR}a(s,\underline{z}) = Y \;\forall\underline{Y} \Rightarrow a(s,\underline{z}) = t(s,\underline{z})$ as above with probability 1. They further proved Theorem 7.3.

Theorem 7.3:

In the class of all pR-unbiased estimators for Y there does not exist one based on $(s, z_i \;/i \in s)$ with the least variance uniformly in $\underline{Y}$.

Writing $z_i^* = z_i - (1-C)\mu_x$ and $t^* = t^*(s,\underline{z}) = 1/C\sum_{i\in s} z_i^*/\pi_i$, the derived HT estimator based on RR data corresponding to the DR-based "open" survey data $(s, y_i | i \in s)$ as derived is

$$t_H = \sum_{i\in s} \frac{z_i}{\pi_i}$$

Further, Adhikary et al. (1984) proved Theorem 7.4.

Theorem 7.4:

For any unbiased estimator $a(s, \underline{z})$ based on $(s, z_i | i \in s)$ for Y,

$$V\left[a(s,\underline{z})\right] = V_p\left[E_R a(s,\underline{z})\right] + E_p\left[V_R\left(t^*(s,\underline{z})\right)\right] + E_p\left[V_R\{a(s,\underline{z}) - t^*(s,\underline{z})\}\right].$$

Hence, they established Theorem 7.5.

Theorem 7.5:

For a pR-unbiased $a(s, \underline{z})$ of Y, if

$$E_m\left[V_{pR}\left(E_R a(s,\underline{z})\right)\right] \geq E_m V_p\left(t_H\right), \text{ then,}$$
$$E_m\left[V_{pR}\left(a(s,\underline{z})\right)\right] \geq E_m V_{pR}\left(t^*(s,\underline{z})\right).$$

Note: $t^*(s, \underline{z})$ thus has a model-based optimality property as possessed by the HT estimator based on an open survey.

They also have Corollary 7.1.

Corollary 7.1:

If a strategy $(p, t_{HT}(s, \underline{Y}))$ has an optimum model-based property so does the strategy $(p, t^*(s, \underline{z}))$ based on $(s, z_i|i \in s)$.

The RR-related results are discussed briefly till now as the concerned details already occur in three published documents by Adhikary et al. (1984), Chaudhuri (1987) and Chaudhuri and Mukerjee (CM, 1988). A later publication by Sengupta and Kundu (1989) is of our next interest in this context and we intend to concentrate on it now.

Drawing mainly upon the preceding works of Eriksson (1973), Adhikary et al. (1984), Chaudhuri (1987) and of course Warner (1965), Sengupta and Kundu (1989) considered RRs $\underline{Z} = (z_1, \ldots, z_i, \ldots, z_N)$ as linear functions of variate values $\underline{Y} = (y_1, \ldots, y_i, \ldots, y_N)$ and linear transform vectors $\underline{R} = (r_1, \ldots, r_i, \ldots, r_N)$ with $E_R(r_i) = y_i$ and $V_R(r_i)$ as a quadratic function of y_is with known coefficients. They considered the following four classes of estimators based on $\underline{Z}$ for Y, namely

$$C_1 = \left\{e(s,\underline{Z}): E_R E_p\left(e(s,\underline{Z})\right) = Y, \forall \underline{Y}\right\}$$

$$C_2 = \left\{e(s,\underline{R}): E_p\left(e(s,\underline{R})\right) = R = \sum_1^N r_i \ \forall \underline{R}\right\}$$

$$C_3 = \left\{e(s,\underline{Z}) = b_s + \sum_{i \in s} b_{si} z_i \ or \ e'(s,\underline{R}) = b'_s + \sum_{i \in s} b'_{si} r_i : E_p\left(b_s\right) = 0 = E_p\left(b'_s\right), \right.$$

$$\left. \sum_{s \ni i} b_{si} p(s) = 1 = \sum_{s \ni i} b'_{si} p(s) \ \forall i \right\}$$

and

$$C_4 = \left\{ e(s,\underline{R}) = \sum_{i \in s} b_{si} r_i, \sum_{s \ni i} b_{si} p(s) = 1 \; \forall i \right\}$$

and observed that $C_1 \supseteq C_2 \supseteq C_3 \supseteq C_4$.

For the classes of designs p_n with a fixed number of n distinct units, p_n^* with n as the average number of distinct units and $p_n(w)$ as the class with $\pi_i = nw_i/W$, $i \in U$ they considered strategies H of (p, e), H_{in}^* of (p_n^*, e), and H_{in}^* of (p_n^*, e), $i = 1, 2, 3, 4$.

Extending an approach of Rao and Bellhouse (1978) of applying C.R. Rao's (1952) theorem these two authors proved the nonexistence of a "uniformly minimum variance" (UMV) estimator in C_i, $i = 1, 2, 3$, for Y confirming the same conclusion essentially reached earlier by Adhikary et al. (1984) and Chaudhuri (1987) and also that in class C_4 the estimator $e_{HT}(s, R) = \sum_{i \in s} r_i/\pi_i$ derived from HT (1952) estimator $\sum_{i \in s} y_i/\pi_i$ in an open survey is the unique UMV for Y when a UCSD is employed, extending to RR surveys the basic result of Hege (1965), Hanurav (1966), and Lanke (1975a). A basic result of Sengupta and Kundu (1989) is

---

Theorem 7.6 (Sengupta and Kundu):

If

$$E_p\big(e(s,\underline{Y})\big) = Y = E_{p'}\big(e'(s,\underline{Y})\big) \quad \text{and} \quad V_p\big(e(s,\underline{Y})\big) \leq V_{p'}\big(e'(s,\underline{Y})\big) \quad \forall \; \underline{Y}, \quad (7.27)$$

then $V_p(e(s,\underline{R})) \leq V_{p'}(e'(s,\underline{R})) \; \forall \underline{Y}$.

If Equation 7.27 is strict over a subset of $\underline{Y}$, then Equation 7.27 is also strict if $\underline{R}$ also belongs to the same subset with a positive probability.

This is a crucial theorem providing corresponding optimality results for estimation and strategies in RR surveys as consequences of results for open surveys. They further affirm that an estimator or a strategy found to be inadmissible in an open survey renders the corresponding derived one for an RR survey as inadmissible too. Arnab's (1995) contention refuting this seems untenable to us. But these authors, however, illustrate one situation when a derived strategy turns out inadmissible when based on RR survey data even though the basic one based on the open survey corresponding to it is itself admissible.

Below are the two lemmas and a theorem they present (ibid).

Lemma 7.1:

For a given strategy $(p, e(s, r))$ in H_2.

$$V\left(e(s, \underline{R})\right) \geq V_p\left[E_R\left(e(s, \underline{R})\right)\right] + \sum_1^N \frac{V_i}{\pi_i}$$

and strictly so if $V_R[e(s, \underline{R}) - e_{HT}(s, \underline{R})] > 0 \quad \forall s$ with $p(s) > 0$.

Lemma 7.2:

If α is a postulated prior distribution of $\underline{Y}$ for which y_is are independently distributed and E_α, V_α are expectation, variance operators with respect to this α, then for a strategy $(p, e(s, \underline{Y}))$ such that $E_p(e(s, \underline{Y})) = Y$, it follows that

$$E_\alpha V_p\left(e(s, \underline{Y})\right) \geq \sum_1^N \frac{V_\alpha(y_i)}{\pi_i} - V_\alpha(Y).$$

An equality follows if

1. $V_\alpha[(e(s, \underline{Y}) - e_{HT}(s, \underline{Y})] = 0$
2. $E_\alpha[e(s, \underline{Y})]$ is a constant for every s with $p(s) > 0$ [cf. Adhikary et al. (1984) for Lemma 7.1 and Godambe and Joshi (1965) for Lemma 7.2]

Theorem 7.7:

Comparing Adhikary et al.'s (1984) "completeness" result $E_R[h(\underline{Z})] = 0$ $\forall \underline{Y} \Rightarrow h(\underline{Z}) = 0$ for every real function; any RR device R for which this is true is denoted as R_c by Sengupta and Kundu (1989).

Basu's (1971) GDE for Y is $\sum_{i \in s}(y_i - \alpha_i)/\pi_i + \sum_1^N \alpha_i$ and a derived GDE for an R or R_c is $(S, \underline{R}) = \sum_{i \in s}(n - \alpha_i)/\pi_i + \sum_1^N \alpha_i$.

An $e_{GD}(s, \underline{R})$ based on R (R_c) is admissible in $C_2(C_1)$ for a given p and a strategy $(p_n, e_{GD}(s, R))$ is admissible in $H^*_{2n}(H^*_{1n})$.

More results relating to the optimality of RR-based results are discussed by them as deduced from corresponding DR-based results, a few of which on modifying some earlier ones by Adhikary et al. (1984) and Chaudhuri (1987), are as follows:

Let M denote a class of probability distributions α of $\underline{Y}$, $\underline{k} = (k_1, ..., k_i, ..., k_N)$, $\underline{w} = (w_1, ..., w_i, ..., w_N)$, $w_i > 0$, $\sum_1 w_i = N$ be known vectors and $x_i = (y_i - k_i)/w_i$, $i \in U = (1, ..., N)$. Let four particular forms of M be

1. M_1 $(\underline{k}, \underline{w})$, such that, x_is have common unknown mean, variance, and covariances.

2. M_2 ($\underline{k}$, $\underline{w}$), such that $x_i s$ are independent with a common unknown mean and a variance.

3. M_3 ($\underline{k}$, $\underline{w}$), such that $\underline{x} = (x_1, ..., x_N)$ has an exchangeable discrete distribution with a common form for each permutation of the subscripts $(i_1, ..., i_N)$ of the coordinates of the $\underline{x}$-vector.

4. M_4 ($\underline{k}$, $\underline{w}$), such that $x_i s$ have an absolutely continuous exchangeable distribution.

They illustrate a situation to show that though a strategy with $e_{HT}(s, \underline{Y})$ is optimal under $M_3(\underline{k}, \underline{w})$, the RR-version $e_{HR}(s, \underline{R})$ is not an element of an optimal strategy. This invalidates Adhikary et al.'s (1984) result in their Corollary 4.1 claiming that the optimality of $e_{HT}(s, \underline{Y})$ implies that of $e_{HT}(s, R)$ under general models M.

Yet Lemma 7.1 of Sengupta et al. (1989) leads to the following.

Theorem 7.8:

If $(p, e_{GD}(s, \underline{Y}))$ be optimal under an M in a subclass of unbiased strategies, then for a given $R(R_c)$

$$E_\alpha \left[V_{Rp'} \left\{ e'(s, \underline{R}) \right\} \right] - E_\alpha \left[V_{Rp} \left\{ e_{GD}(s, \underline{R}') \right\} \right] \geq 0 \text{ for every } \alpha \text{ in } M;$$

here $(p', e'(s, R))$ belongs to the corresponding subclass of $H_2(H_1)$ with $\pi_i(p') = \pi_i(p)$ $\forall i$. Here $V_{RP'}$ = variance operator for an RR device and a sampling design p' while V_{Rp} is that with p' replaced by design p.

Sengupta et al. (1989) affirm that from the known optimality results in the open setup [cf. Cassel et al. (1976), Godambe and Joshi (1965), Godambe and Thompson (1973, 1977)], one may establish the optimality of a $(p_n(w), e_{GD}(s, \underline{R}))$ in the subclass of H_{3n} under $M_1(\underline{k}, \underline{w})$ and in H_{2n} (or H_{1n} for an R_c) under $M_i(\underline{k}, \underline{w})$, $i = 2, 3, 4$. It follows that two theorems in Chaudhuri (1987) follow as consequences and another one is invalidated unless p_n is restricted to $p_n(w)$ in view of these comprehensive studies by Sengupta and Kundu (1989), which also contains further results of minor importance to which every reader's attention is drawn.

Thus, we remain confined to the coverage of unbiased estimators alone. In terms of the accuracy achievable it behoves us to pay attention to biased estimators as well especially if ADU may be ensured for an estimator for Y employed. Chaudhuri et al. (1996) started with the RR set up with $\underline{R} = (r_1, ..., r_i, ..., r_N)$, $E_R(r_i) = y_i$, r_i's independent, $V_R(r_i) = \alpha_i y_i^2 + \beta_i y_i + \theta_i = V_i$, $i \in U = (1, ..., i, ..., N)$ with known $\alpha_i, \beta_i, \theta_i$, $(1 + \alpha_i) \neq 0$, $v_i = 1/(1 + \alpha_i)(\alpha_i r_i^2 + \beta_i r_i + \theta_i)$, $E_R(v_i) = V_i$ and postulated a model for $\underline{Y} = (y_1, ..., y_i, ..., y_N)$. So, they could write $y_i = \beta x_i + \epsilon_i$, β an unknown constant, $x_i s$ as positive known numbers,

$X = \Sigma_1^N x_i$ and $\in_i$s as independent random variables with model moments $E_m(\in_i) = 0$, $V_m(\in_i) = \sigma^2 x_i^g$, $\sigma(>0)$ unknown, $0 \le g \le 2$ but g unknown otherwise. By $\Sigma, \Sigma\Sigma$ denoting sums over $i = 1, ..., N$ and $i < j$ over 1 through N and by $\Sigma', \Sigma'\Sigma'$ the same over units in a sample s and paired units in s with no duplication, for $Y = \Sigma y_i$ they considered the derived Greg estimator

$$e_g = \sum{}' \frac{r_i}{\pi_i} + \hat{\beta}_{Qr}\left(X - \sum{}' \frac{x_i}{\pi_i} \right) = \sum{}' \frac{r_i}{\pi_i} g_{si},$$

writing

$$g_{si} = 1 + \left(X - \sum{}' \frac{x_i}{\pi_i} \right) \frac{x_i Q_i \pi_i}{\sum{}' x_i^2 Q_i}, \quad \hat{\beta}_{Qr} = \frac{\sum{}' r_i x_i Q_i}{\sum{}' x_i^2 Q_i}, \quad Q_i\,(>0,)$$

to be suitably chosen.
   Writing

$$a_{1i} = 1, \quad a_{2i} = g_{si}, \quad \Delta_{ij} = \left(\frac{\pi_i \pi_j - \pi_{ij}}{\pi_{ij}} \right), \quad \pi_{ij} > 0, \quad e_{ir} = r_i - \hat{\beta}_{Qr} x_i,$$

$$v_{kr} = \sum{}'\sum{}' \Delta_{ij} \left(\frac{e_{ir}}{\pi_i} a_{ki} - \frac{e_{jr}}{\pi_j} a_{kj} \right)^2, \quad k = 1, 2$$

they worked out

$$E_R\left(e_{ir} - e_i \right)^2 = V_i + x_i^2 \frac{\sum{}' x_i^2 Q_i V_i}{\left(\sum{}' x_i^2 Q_i \right)^2} - \frac{2 x_i^2 Q_i V_i}{\sum{}' x_i^2 Q_i} = F_i,$$

and

$$E_R\left(e_{ir} - e_i \right)\left(e_{jr} - e_j \right) = -x_i x_j \left[\frac{Q_i V_i + Q_j V_j}{\sum{}' x_i^2 Q_i} - \frac{\sum{}' x_i^2 Q_i^2 V_i}{\left(\sum{}' x_i^2 Q_i \right)^2} \right] = F_{ij},$$

say and proposed

$$v_{kg} = v_{kr} - \sum{}'\sum{}' \Delta_{ij} \left[\left(\frac{a_{ki}}{\pi_i} \right)^2 \hat{F}_i + \left(\frac{a_{kj}}{\pi_j} \right)^2 \hat{F}_j - 2 \frac{a_{ki} a_{kj}}{\pi_i \pi_j} \hat{F}_{ij} \right] + \sum{}' \left(\frac{g_{si}}{\pi_i} \right)^2 \hat{V}_i$$

for $k = 1, 2$ as two approximate variance estimators for e_g.

Alternatively, approximating $Ep(e_g)$ by $R = \sum r_i$ and $M = E_p E_R (e_g - Y)^2$ by $E_R V_p(e_g) + \sum V_i$ they approximated $V_p(e_g)$ in the following way. To achieve this they introduced the following notations:

$$w_{1i} = r_i, \quad w_{2i} = x_i, \quad w_{3i} = r_i x_i Q_i \pi_i, \quad w_{4i} = x_i^2 Q_i \pi_i,$$

$$\hat{T}_j = \sum{}' \frac{w_{ji}}{\pi_i}, \quad T_j = \sum w_{ji} \ (j = 1, \dots, 4)$$

$$\hat{\underline{T}} = \left(\hat{T}_1, \dots, \hat{T}_4 \right), \quad \underline{T} = \left(T_1, \dots, T_4 \right)$$

leading to

$$e_g = \hat{T}_1 + \frac{\hat{T}_3}{\hat{T}_4} \left(X - \hat{T}_2 \right) = f\left(\hat{\underline{T}} \right), \text{ say.}$$

Then

$$e_g = f\left(\underline{T} \right) + \sum_{j=1}^{4} \lambda_j \left(\hat{T}_j - T_j \right), \quad \lambda_j = \frac{\delta f\left(\hat{\underline{T}} \right)}{\delta \hat{T}_j} \bigg|_{\hat{\underline{T}} = \underline{T}}.$$

Let

$$\phi_i = \sum_{j=1}^{4} \lambda_j w_{ji}, \quad \hat{\lambda}_j = \lambda_j \bigg|_{\underline{T} = \hat{\underline{T}}}, \quad \hat{\phi}_i = \sum_{j=1}^{4} \hat{\lambda}_j w_{ji}, \quad V_p\left(e_g \right) \text{ by } V_p\left(\sum{}' \frac{\phi_i}{\pi_i} \right),$$

and two additional variance estimators follow as

$$m_{1g} = \sum{}' \sum{}' \Delta_{ij} \left(\frac{\hat{\phi}_i}{\pi_i} - \frac{\hat{\phi}_j}{\pi_j} \right)^2 + \sum{}' \frac{\hat{V}_i}{\pi_i}$$

and

$$m_{2g} = \sum{}' \frac{\hat{\phi}_i^2}{\pi_i} \left(\frac{1}{\pi_i} - 1 \right) + \sum{}' \sum{}' \frac{\hat{\phi}_i \hat{\phi}_j}{\pi_{ij}} \left(\frac{\pi_{ij} - \pi_i \pi_j}{\pi_i \pi_j} \right) + \sum{}' \frac{\hat{V}_i}{\pi_i}.$$

Chaudhuri et al. (1998) considered the following three alternative forms of Greg estimators for Y based on RR-based survey data, namely

$$e_1 = \sum{}' \frac{r_i}{\pi_i} + \left(X - \sum{}' \frac{x_i}{\pi_i} \right) b_{Qr}$$

$$e_2 = \sum{}' r_i b_{si} + \left(X - \sum{}' x_i b_{si} \right) b_{Qr}$$

$$e_3 = \sum{}_n r_i \frac{P_i}{p_i} + \left(X - \sum{}_n x_i \frac{P_i}{p_i} p_i \right) b_{Qr}$$

with b_{Qr} same as $\hat{B}_{Qr}$ solved earlier, b_{si} subject to $\sum_{s \ni i} p(s) b_{si} = 1 \; \forall i$ and e_3 is based on, RHC scheme with the associated notations discussed earlier except that here P_i means the sum of the normed size-measures p_i over the units falling in the ith group. Their MSEs are

$$M(e_j) = E_R V_p(e_j) + \sum V_i.$$

Writing

$$B_Q = \frac{\sum r_i x_i Q_i \pi_i}{\sum x_i^2 Q_i \pi_i}, \quad f_i = \frac{p_i}{P_i}, \quad C_Q = \frac{\sum r_i x_i Q_i f_i}{\sum x_i^2 Q_i f_i}, \quad G_i = r_i - B_Q x_i, \quad D_i = r_i - C_Q x_i,$$

they get approximately

$$M(e_1) = E_R V_p \left(\sum \frac{G_i}{\pi_i} I_{si} \right) + \sum V_i$$

$$M(e_2) = E_R V_p \left(\sum G_i b_{si} I_{si} \right) + \sum V_i$$

$$M(e_3) = E_R V_p \left(\sum{}_n D_i \frac{P_i}{p_i} \right) + \sum V_i$$

using Taylor-series approximation as was done by Chaudhuri et al. (1996) as discussed earlier.

Writing $g_i = r_i - b_{Qr} x_i$, choosing some nonzero numbers u_i they take

$$m(e_1) = \sum \sum{}' \Delta_{ij} \left(\frac{g_i}{\pi_i} - \frac{g_j}{\pi_j} \right)^2 + \sum{}' \frac{\hat{V}_i}{\pi_i},$$

$$m(e_2) = - \sum \sum{}' d_{sij} u_i u_j \left(\frac{g_i}{u_i} - \frac{g_j}{u_j} \right)^2 + \sum{}' \hat{V}_i C_{si} I_{si},$$

with obvious notations as appropriate estimators for $M(e_1)$ and $M(e_2)$, respectively. Further writing for the RHC scheme adopted

$$a_{ij}(1) = \frac{\sum_n N_i^2 - N}{N^2 - \sum_n N_i^2} p_i p_j \quad \text{and} \quad a_{ij}(2) = \frac{\sum_n N_i^2 - N}{n(n-1)} \frac{p_i p_j}{N_i N_j},$$

they recommend the following two estimators for $M(e_3)$ adhering to choices by RHC (1962) and Ohlsson (1989), respectively—

$$m_k(e_3) = \sum_n \sum_n a_{ij}(k) \left(\frac{g_i}{p_i} - \frac{g_j}{p_j} \right)^2 + \sum_n \hat{V}_i \frac{P_i}{p_i}, \quad k = 1,2$$

—on utilizing the fact that

$$v_k = \sum_n \sum_n a_{ij}(k) \left(\frac{y_i}{p_i} - \frac{y_j}{p_j} \right)^2$$

has

$$E_p(v_k) = V_p \left(\sum_n y_i \frac{P_i}{p_i} \right), \quad k = 1, 2.$$

In the context of estimating variances and MSEs of estimators for Y from RR-based survey data, other references recommended for curious readers are Chaudhuri (1992, 1993) and one more by Chaudhuri and Adhikary (1989). Arnab's (1995, 1996) context of RR will be discussed in Chapter 9.

Next, let us see how optional randomization technique of Chaudhuri and Dihidar (2009) may work in this context of quantitative characteristics anticipated to carry social stigmas.

Let a person labeled i may, with an unknown probability $C_i (0 \le C_i \le 1)$, choose to give out the true response y_i or with the complementary probability $(1 - C_i)$ opt to give an RR following an RR device proffered to the person. Let the RR device's stipulated demand, drawing at random from a box one of the cards that are marked with numbers $a_1, ..., a_j, ..., a_m$ with a mean $\mu_a = (1/m)\sum_1^M a_j = 1$ and independently and randomly one of the cards bearing marks $b_1, ..., b_L$ with a mean $\mu_b = (1/L)\sum_1^L b_k$ and the person be also required to draw randomly and independently from a third box with cards bearing the numbers $b_1', ..., b_L'$ with $\mu_b' = (1/L) \sum_1^L b_i' \ne \mu_b$. The box from which the respective cards are picked is not to be disclosed to the enquirer by the respondent. With probability $(1 - C_i)$ the two RRs from the ith person will be of the form

$$I_i = a_j y_i + b_k \quad \text{and} \quad I'_i = a_b y_i + b'_\mu, \ \text{say.}$$

Then, let $z_i = y_i$ with probability C_i
    and $= I_i$ with probability $(1 - C_i)$ and independently, let
    $z'_i = y_i$ with probability C_i
        $= I'_i$ with probability $(1 - C_i)$

Then one may calculate

$$E_R\left(z_i\right) = C_i y_i + \left(1 - C_i\right)\left[y_i \mu_a + \mu_b\right] = C_i y_i + \left(1 - C_i\right)\left(y_i + \mu_b\right)$$

and

$$E_R\left(z'_i\right) = C_i y_i + \left(1 - C_i\right)\left[y_i \mu_a + \mu'_b\right] = C_i y_i + \left(1 - C_i\right)\left(y_i + \mu'_b\right)$$

Then, from $E_R(\mu'_b z_i - \mu_b z'_i) = (\mu'_b - \mu_b)y_i$ one may find $r_{1i} = (\mu'_b z_i - \mu_b z'_i)/(\mu'_b - \mu_b)$ with $E_R(r_i) = y_i$.
  Repeating the entire exercise independently once again another such independent unbiased estimator r_{2i} with $E_R(r_{2i})$ may be generated to yield

$$r_i = \frac{1}{2}(r_{1i} + r_{2i}) \text{ with } E_R(r_i) = y_i$$

and $v_i = 1/4(r_{1i} - r_{2i})^2$ with $E_R(v_i) = V_i = V_R(r_i)$. Then, as usual, unbiased estimators $e = \sum r_i b_{si} I_{si}$ or ADU like $e_g = \sum r_i g_{si} I_{si}$ may be constructed for Y along with variance or MSE estimators already described in detail so far.
  As the literature has so far not discussed about how privacy may be protected once quantitative RRs compulsorily or optionally are given out, we need not report anything here on this important subject.

8

Other Indirect Questioning Techniques

Introduction

Some social scientists are of the opinion that RRTs are too exacting in their applicability from the intended respondents. As a possible alternative they prefer an item count technique (ICT). In this, a certain number G of innocuous items are incorporated in a structured questionnaire asking a sampled person in a chosen s_1 sample as to how many of them are applicable to the person. Another sample s_2 is drawn independently of the former sample s_1 and a person in it is offered a questionnaire with the same G items as the person in the first sample s_1.

However, most importantly, a $(G + 1)$st item is incorporated into the questionnaire for the first sample s_1, in addition to asking the sampled person's bearing a "tainted" feature T or a "fresh" feature F or both T and F while a $(G + 1)$st item is inserted in the questionnaire addressed to a person in the second sample s_2, enquiring about the person's bearing either the complement of T as T^c or the complement of F as F^c or both T^c and F^c. But the response demanded of a person in s_1 is just the number out of $G + 1$ items in the questionnaire applicable to the person and exactly a similar one for the person addressed in the sample s_2. If the proportion bearing the innocuous feature F is assumed as known, then it is easy to estimate the proportion bearing T from the above survey data with secrecy well protected concerning the "taint."

The nominative technique recommended by several social scientists is a device that avoids asking the sampled person about the person bearing the stigmatizing characteristic of oneself, but only if the person is aware of it being true of some of the person's acquaintances with no particular references given out. This is analogous to "Network" sampling, if necessary, combined with adaptive sampling applicable in tandem.

Another alternative is known as "The Three-Card" method. This uses three boxes with varying assortment of cards with items of enquiry inscribed thereupon containing the sensitive item and a few innocuous ones ingeniously worded to induce truthful responses with a convincing protective apparatus inbuilt to retain the jeopardizing features securely hidden.

Details of these three alternatives to RRT are now set forth in the three succeeding sections.

Item Count Technique

Droitcour et al. (1991) introduced the Item Count Technique (ICT) as a means to procure data on sensitive characteristics by dint of indirect queries. Chaudhuri and Christofides (2007) provided an amendment which we now recount in brief.

Let θ denote as usual the unknown proportion of people bearing a "tainted" feature T in a community of persons. Suppose an enquirer needs an unbiased estimator $\hat{\theta}$ on surveying people from the community by indirect procedures and avoiding direct questioning because of suspecting popular disapproval and tendency for reluctance to divulge truths.

Suppose two independent probability samples s_1 and s_2 of suitable sizes are taken following a common sampling design with inclusion probabilities π_i for a unit i of the population $U = (1, ..., i, ..., N)$ and π_{ij} for distinct pairs of units (i, j) of U. To every person in s_1, let a questionnaire with $(G + 1)$ listed items of enquiry be presented such that G of the items are innocuous commonly applicable or otherwise in people, but one item needs to gather if the person bears the tainted feature T or a fresh nonoffensive feature F or both T and F in conjunction with one or more preceding G items, the addressee is to give out the "number out of these $G + 1$ items" applicable to him/her.

On the other hand, to every person in the sample s_2, let a questionnaire be offered listing the same G innocuous items as earlier plus another one asking for the person's bearing T^c the complement of T, or F^c the complement of F, or $T^c \cap F^c$. To every person in s_1 as well as to every one in s_2, the question asked is just "the number out of $(G + 1)$ items thus listed" that actually applies to the person. This leads to the following developments:

$$Nt_1 = t(s_1) = \sum_{i \in s_1} \frac{y_i}{\pi_i}, \quad Nt_2 = t(s_2) = \sum_{j \in s_2} \frac{x_j}{\pi_j},$$

y_i = number reported by i of s_1
x_j = number reported by j of s_2.

Let θ_F be the known proportion out of the N persons bearing F, say, a known proportion of those who completed school education or the known proportion of those who are government employees, and so on. Hence, we get

---

Theorem 8.1:

$\hat{\theta} = (t_1 - t_2) + (1 - \theta_F)$ is an unbiased estimator of θ.

Proof: $E_p(t_1) =$ Proportion bearing $(T \cup F)$ in conjunction with one or more or less of the G items in the community

= Proportion bearing T
  + Proportion bearing F
  − Proportion bearing $T \cap F$

$E_p(t_2)$ = Proportion in the community bearing $T^c \cup F^c$ in conjunction with none or one or more of the G innocuous items.

= 1 – Proportion bearing $T \cap F$ using DeMorgan's Law giving $T^c \cup F^c = (T \cap F)^c$.

So, $E_p(t_1) - E_p(t_2) = \theta + \theta_F - 1$.

So, $E_p(\hat{\theta}) = \theta$. Utilizing the results for variance of the Horvitz and Thompson's estimator for a finite population total and of its unbiased variance estimator we get:

Theorem 8.2:

1. $V_p(\hat{\theta}) = \dfrac{1}{N^2}\left[\displaystyle\sum_{k=1}^{N}\sum_{k+1}^{N}(\pi_k\pi_e - \pi_{ke})\left\{\left(\dfrac{y_k}{\pi_k}-\dfrac{y_l}{\pi_l}\right)^2 + \left(\dfrac{x_k}{\pi_k}-\dfrac{x_l}{\pi_l}\right)^2\right\}\right.$

$\left. + \displaystyle\sum_{k=1}^{N}\dfrac{\beta_k}{\pi_k}\left(y_k^2 + x_k^2\right)\right]$

is the variance of $\hat{\theta}$, writing

$$\beta_k = 1 + \dfrac{1}{\pi_k}\sum_{l=1,l\neq k}^{N}\pi_{kl} - \sum_{1}^{N}\pi_k$$

2. $v_p(\hat{\theta}) = \dfrac{1}{N^2}\left[\displaystyle\sum_{k\in s_1}\sum_{l\in s_2}\left(\dfrac{\pi_k\pi_l - \pi_{kl}}{\pi_{kl}}\right)\left(\dfrac{y_k}{\pi_k}-\dfrac{y_l}{\pi_l}\right)^2 + \sum_{k\in s_2}\sum_{l\in s_2}^{l>k}\left(\dfrac{\pi_k\pi_l - \pi_{kl}}{\pi_{kl}}\right)\right.$

$\left. \times \left(\dfrac{x_k}{\pi_k}-\dfrac{x_l}{\pi_l}\right)^2 + \displaystyle\sum_{k\in s_1}\dfrac{y_k^2}{\pi_k^2}\beta_k + \sum_{k\in s_2}^{\beta_k>k}\dfrac{x_k^2}{\pi_k^2}\beta_k\right]$

assuming $\pi_{kl} > 0 \ \forall \ k \neq l$, is an unbiased estimator for $V_p(\hat{\theta})$.

Questionnaire for s_1

Item	Description
1	I was never hospitalized
2	I do have hay fever
3	I have no eczema
4	I took antibiotics in the last 2 years
5	I consume marijuana and I have asthma
6	My father smokes cigarettes
7	I consider smoking harmful

Questionnaire for s_2

Item	Description
1	I was never hospitalized
2	I do have hay fever
3	I have no eczema
4	I took antibiotics in the last 2 years
5	I never consumed marijuana and I never had asthma
6	My father smokes cigarettes
7	I consider smoking harmful

Quoting from Chaudhuri and Christofides's (2007) two questionnaires, respectively for respondents from s_1 and from s_2 to illustrate how an ICT works in practice.

Here $G = 6$, consuming marijuana is T and "having asthma" is F with a known incidence/prevalence rate from the current/recent medical report.

Nominative Technique

Miller (1985) introduced the "nominative technique" in the context of indirect questioning to elicit truthful responses relating to stigmatizing characteristics purported to ensure protection of privacy.

As usual, A is a stigmatizing attribute and we need to estimate θ, the proportion of people bearing it in a community of persons identified in a labeled population $U = (1, ..., i, ..., N)$ by dint of a suitable survey, gathering data not by direct questioning but in an indirect manner as described below.

From the population, a sample s of a number of people is selected on following an appropriate design p with π_i, π_{ij} denoting inclusion probabilities of individuals i and of paired individuals i, $j(i \neq j)$, all assumed positive and further, $\beta_i = 1 + (1/\pi_i)\sum_{j=1 \neq i}^N \pi_{ij} - \sum_1^N \pi_i$.

A person j in s when selected is requested to truthfully identify, giving particulars of age, sex, marital status, occupation, level of education, family size, residential location, but not the actual address of each of the person's close acquaintances among the persons in U about whom the person is sure that the latter bears the characteristic A. The person j who is selected according to the design is the "informant" analogous to what is called technically the "selection unit" (SU) in "Network" sampling. The named person is, labeled as i in U is the person's "nominee." This nominee is like the "observation unit" (OU) in network sampling. But unlike an OU in network sampling the "nominee" in the "nominative technique" is not further sought for the person's whereabouts or any particulars for interrogation. The "informant" is simply trusted. Demographic and social features of the nominees are gathered from the informant(s) only to develop estimates for the "number and proportion" of persons in the community bearing the sensitive characteristic A classified by various socioeconomic-demographic characteristics denoted by C.

Let a possible nominee of the jth SU be one who is technically said to be "linked" to the nominee and μ_j be the "set" of such persons in U "linked" to the jth possible informant. Let m_i be the total "number" of people in U who are aware that the ith person in U bears the stigmatizing feature A and who may, if approached, report so as a possible informant. This number m_i is sought to be accurately gathered only by contacting the sampled informants.

Let

$$I_i(A) = 1 \quad \text{if } i \text{ bears } A$$
$$= 0 \quad \text{if } i \text{ does not bear } A.$$

Let $w_j = \sum_{i \in \mu_j} I_i(A)/m_i$ and N_A be the total number of unknown persons in U bearing A. It follows that $\sum_{j=1}^{N} w_j = N_A$.

Our objective is to simply estimate N_A and equivalently $\sum_{j=1}^{N} w_j$ using the information supplied by the sample of "informants" giving the set of nominees "named." So, our proposed estimator for N_A is

$$e = \sum_{j \in s} \frac{w_j}{\pi_j}$$

An unbiased variance-estimator for e by Chaudhuri and Pal's (2002) result is

$$v_p(e) = \sum_{k \in s} \sum_{\substack{l \in s \\ l > k}} \frac{(\pi_k \pi_l - \pi_{kl})}{\pi_{kl}} \left(\frac{w_k}{\pi_k} - \frac{w_l}{\pi_l} \right)^2 + \sum_{k \in s} \frac{w_k^2}{\pi_k} \frac{\beta_k}{\pi_k}$$

One important feature which is an advantage over network sampling in this nominative technique is that we need not bother about the increasing magnitude of m_i and of the extent of the volume of μ_j because there is no plan to survey the nominees.

A second feature that is noteworthy is that if the selected sample s does not yield too many informants then only a negligible fraction of N_A may be referred to and accounted for. So, the nominative technique has been supplemented in the following way by Chaudhuri and Christofides (2008).

They recommend moving ahead from the initially chosen sample s to an adaptive sample $A(s)$ on adopting "adaptive sampling," a well-known device in survey sampling in augmenting the information content of a primary sample as authentically described by Chaudhuri (2000). Also, they recommend supplementing the technique further by resorting to the tactic of Chaudhuri et al. (2004) to having an inbuilt method of putting a brake if there is an excessive growth in the size of $A(s)$ calling for escalated budget provisions.

For every "su" j in U let there be a uniquely defined neighborhood composed of j plus those of the informant's close acquaintances among the units of U who reciprocally may include j, if requested to mention their close acquaintances among the units of the same U. If a sampled person j nominates at least one known person in U to bear A, then every person in the neighborhood qualifies to be approached for nominating the latter's acquaintances believed to bear A; this practice is similarly to continue till the last person in the growing neighborhood fails to nominate any. The collection of the units of U thus covered starting with the originating unit j constitutes the "cluster" of the jth person. Omitting from this cluster the individuals who on being approached decline to nominate a single A-bearing person, the left-over set is called the "network" of j, denoted $n(j)$. Any person in U not bearing A and not nominating any person bearing A is named a "Singleton network." Courteously recognizing the "Singleton networks" as genuine "networks" it follows that U is the union of all the "networks," each of which is clearly "disjoint" from every other. Denoting by C_j the cardinality of $n(j)$, that is, the number of units in $n(j)$, it follows that $a_j = (1/C_j)\sum_{k\in n(j)} w_k$ satisfies the condition

$$\sum_{j=1}^{N} a_j = \sum_{k=1}^{N} w_k = W = N_A.$$

Hence, $t = \sum_{j\in s} a_j/\pi_j$ is an unbiased estimator for N_A. Most importantly, one needs to note that in order to ascertain the values of a_j for $j \in s$ one has to survey all the units in $U_{j\in s}n(j)$, which constitutes the "Adaptive Sample" $A(s)$ emanated from the initial sample s. Hopefully, this procedure substantially promotes the information content of the survey data produced by $A(s)$ affording a reasonably accurate estimate of N_A. If, however, $A(s)$ is too inflated vis-à-vis U then one may apply Chaudhuri et al.'s (2004) "Restricted Adaptive" procedure. This entails subsampling of $b_j(<C_j)$ units out of the respective $n(j)$ say, independently across j in s yielding data from subsets $m(j)$ out $n(j)$'s choosing b_j's such that $\sum_{j\in s} b_j \le B$, say, so that one may afford to survey upto B units of U to find an effective set and number of informants. Then

$$d_j = \frac{C_j}{b_j} \sum_{k\in m(j)} w_k$$

is unbiased for a_j independently over j in s and

$$e = \sum_{j\in s} \frac{d_j}{\pi_j}$$

provides an unbiased estimator for $N_A = W$.

If we need to estimate N_A by categories say $N_A(C)$, using C for age, sex, occupation, residential specialties, education, and so on we have to replace I_i (A) by I_i (A/C) in the formula throughout, writing

I_i $(A/C) = 1$ if the ith person of category C bears A
     $= 0$ if such a person does not bear A.

Using results discussed earlier, it follows that

$$\hat{V}(t) = \sum_{j<}\sum_{k\in s}\left(\frac{\pi_j\pi_k - \pi_{jk}}{\pi_{jk}}\right)\left(\frac{a_j}{a_j} - \frac{a_k}{\pi_k}\right)^2 + \sum_{k\in s}\frac{\beta_k^2}{\pi_k^2}I_i(A)$$

and

$$\hat{V}(e) = \hat{V}(t)\bigg|_{\substack{a_j = d_j \\ j\in s_j}} + \sum_{j\in s}\frac{1}{\pi_j}\frac{\left[(1/b_j - 1/C_j)\right]}{(b_j - 1)}\left(\sum_{k\in m_j}w_j - \frac{\sum_{k\in m(j)}w_k}{b_j}\right)^2\right).$$

Three-Cards Method

Droitcour et al. (2001) gave this indirect method of data gathering about a sensitive characteristic A. Chaudhuri and Christofides (2008) briefly recounted it. The present one is an adaptation from both as per our absorption on their perusal.

The population is supposed to be divided into four mutually exclusive but not exhaustive groups A, B, C, and D of which only group A carries the stigmatizing feature, the other three being innocuous. The groups in practice are clearly described in elucidating language narrated in the "cards." To gather the data, three independent samples are taken and for every sampled person three boxes are used in which are placed suitably worded cards as follows. Hence the "nomenclature" for the method.

Card Type

 Sample 1: Box 1: I am of B-type.

  Box 2: I am of C or D or A-type.

  Box 3: I am of a group not A, B, C, or D.

 Sample 2: Box 1: I am of C-type.

  Box 2: I am of B or D or A-type.

  Box 3: I am different from A, B, C, D.

 Sample 3: Box 1: I am of D-type.

Box 2: I am of B or C or A-type.

Box 3: I am not of A, B, C, D-type.

Every sampled person is to give out only the box number that applies to self.

Writing θ_A, θ_B, θ_C, θ_D, θ_E for the proportions of U in the groups A, B, C, D, and the remaining section of U, respectively, unbiased estimators $\hat{\theta}_B$, $\hat{\theta}_C$, $\hat{\theta}_D$ for θ_B, θ_C, and θ_D are directly available from the samples s_1, s_2, s_3, say, namely the first three samples, respectively.

Let $\hat{\theta}_{1ACD}$ be an unbiased estimator obtained from the first sample s_1 of the proportions choosing the second box, then

$$\hat{\theta}_{1A} = \hat{\theta}_{1ACD} - \hat{\theta}_C - \hat{\theta}_D$$

is an estimator (unbiased of course) of the proportion bearing the sensitive attribute A as determined from the first sample s_1. Two other unbiased and independent estimators for θ_A are

$$\hat{\theta}_{2A} = \hat{\theta}_{2ABD} - \hat{\theta}_B - \hat{\theta}_D \text{ from } s_2$$

and

$$\hat{\theta}_{3A} = \hat{\theta}_{3BCA} - \hat{\theta}_C - \hat{\theta}_B \text{ from } s_3$$

$\text{Var}(\hat{\theta}_{1A})$, $\text{Var}(\hat{\theta}_{2A})$, $\text{Var}(\hat{\theta}_{3A})$, and their unbiased estimators $\hat{\text{Var}}(\hat{\theta}_{1A})$, $\hat{\text{Var}}(\hat{\theta}_{2A})$, and $\hat{\text{Var}}(\hat{\theta}_{3A})$ follow easily noting the independence of the samples and the generality of the sampling designs that may be employed.

So far, one common shortcoming of the three indirect techniques discussed in this chapter under "Item Count Technique," "Nominative Technique," and "The Three-Cards Method" is that they are thus far applicable only to qualitative characteristics. No literature so far is available to throw light on how alternatives to the RR technique may be developed to cover quantitative characteristics.

9

Miscellaneous Techniques, Applications, and Conclusions

Introduction

Sinha and Hedayat (1991) presented their thoughts in their book. Sarjinder Singh (1996), in his book as well as in numerous research papers, mainly collaborating with Mangat Singh, Late Ravindra Singh, Late Tracy, Joarder, Horn, Gjestvang and others, extensively dealt with RR literature. N.S. Mangat, in collaboration with others, gave us relevant material that we need to dwell on. Raghunath Arnab, Tapan Nayak, Kim and Warde, and quite a few other prolific contributors deserve our attention in giving us a substantial growth in the subject we have undertaken to review in this treatise. A condensed description plus a critical appreciation is modestly attempted below.

Review

Sinha and Hedayat

Echoing the voice of many other workers in the area of randomized response (RR), Sinha and Hedayat (1991, p. 310) also announce that data gathering on sensitive issues "calls for special 'sampling' techniques which ensure a high degree of confidentiality to the respondents." We believe we have by now succeeded in putting across our view to the contrary that it is enough to try any specific RR device on any single individual, "absolutely no matter how sampled." What we need specially is an alternative "survey technique" but no new "sampling technique." This misconception is feared to have been at the root of the traditional practice of tying up the RR devices with the simple random sampling with replacement (SRSWR) scheme alone and thereby keeping the RR literature at a rudimentary level away from the well-developed direct response (DR) sample survey theory at a higher level of scientific status for rather too long.

The resulting growth of RR theory inalienably linked to SRSWR rendered it a great simplicity emanating from the tact that "every randomly chosen person has the unique probability of bearing the tainted attribute A say, equal to θ, the unknown proportion bearing this A in the entire community." In estimating this θ and also in defining an appropriate measure of jeopardy in revealing one's genuine attribute through truthful RR this affords the RR theory a great deal of simplicity indeed.

Sinha and Hedayat (1991) envisaged a population of N persons with the ith ($i = 1, \ldots, N$) person bearing a value y_i, which is either 1 or 0, accordingly as i bears A or its complement A^c, and ϕ as the unknown proportion valued 1 each in the community but stumble on the stochastic formulation that a response from the ith person, namely y_i has the probability distribution

$$P(y_i = 0/\text{if } i \text{ bears } A^c) = 1$$

$$P(y_i = 1/\text{if } i \text{ bears } A^c) = 0$$

$$P(y_i = 1/\text{if } i \text{ bears } A) = q_{i1}$$

$$P(y_i = 0/\text{if } i \text{ bears } A) = q_{i2}$$

and $P(\text{no response elicited}/i \text{ bears } A) = q_{i0}$ so that $q_{i0} + q_{i1} + q_{i2} = 1$, $i = 1, \ldots, N$. Interestingly, they need not bother about the particular draw when the ith person responds. But these authors refrain from sticking to this formulation owing to compulsions for simplicity. Also they presume that anyone bearing A^c has no propensity to suppress this truth if so requested.

In their chapter "Method 1: Related Question Procedure," pp. 312–318, Sinha and Hedayat define

$$\phi_i = 1 \quad \text{if } i \text{ bears } A$$

$$= 0 \quad \text{if } i \text{ bears } A^c$$

and proceed to estimate $\phi = 1/N \sum_1^N$ in their own way with no reference to Warner (1965) or anyone else. They further developed, "Method 2: Unrelated Question Procedure," pp. 319–324 and more techniques dealing with estimation concerning sensitive attributes (pp. 325–330) to which all readers may draw their attention because they are quite different from most RRT dealt with elsewhere. Their developments, however, proceed initially with no connection with any specific sampling design. The literature on RR is incomplete without incorporating the contents of Chapter 11 by Sinha and Hedayat (1991). But in Section 11.4 in their book, which deals with sensitive quantitative variables, is quite intriguing and cannot but draw one's attention to the period of bitter controversies among the celebrated survey sampling experts Hartley and Rao (1968, 1971a,b), on the one hand, or Godambe (1970) on the other. A curious reader may have a look at a summarized version in Chaudhuri and Vos (1988). Sinha and Hedayat (1991, p. 331) assume "that the survey statistician has a set of M distinct values $K_1, ..., K_M$ available and these are meant to cover the whole set of values under study." It seemed queer that these authors did not oppose the theory developed by Adhikary et al. (1984), Chaudhuri (1987), Sengupta and Kundu (1989) as detailed in the Chapter 7 of this book. One who is aware of the "Hartley and Rao versus Godambe controversy" would find it hard to reconcile the material in Section 11.47 of Sinha and Hedayat (1991) with that dealt with in Chapter 7 of this book. As their coverage of the subject contains several intriguing issues, we feel constrained not to attempt any further discussions.

R. Arnab

Arnab (1995a) presented certain modifications on the earlier works by Adhikary et al. (1984), Chaudhuri (1987, 1992), among others mentioned in Chapter 7 but did not deal with those by Sengupta and Kundu (1989). We intend to briefly comment on them. His intention was to provide optimal estimation for a finite population totally under RR surveys. To estimate $Y = \Sigma_1^N y_i$, he allowed a general RR device yielding suitably transformed RRs as r_i independently across the individuals $i \notin U, U = (1, ..., i, ..., N)$ bearing sensitive real values y_i such that

$$E_R(r_i) = y_i, \quad V_R(r_i) = \phi_i(>0) \quad C_R(r_i, r_j) = 0 \ \forall i, j$$

writing as usual E_R, V_R, C_R as expectation, variance, and covariance operators with respect to randomization device adopted and ϕ_i is a known function of y_i's. Denoting a sample by s, chosen with a probability $p(s)$ according to a design p which is further specified by p_n in case every s has a fixed effective size n and by p_n^* if in addition p_n involves the inclusion-probability $\pi_i = \Sigma_{s \ni i} p(s)$, such that $\pi_i = n(x_i/X)$, if $x_i (>0)$ are known size measures well correlated with y_i's and $X = \Sigma_1^N x_i$.

Because of the general nature of ϕ_i, the model envisaged in the approach of Arnab (1995a) is wider than the ones studied by Eriksson (1973), Godambe (1980), Chaudhuri and Adhikary (1981), Adhikary et al. (1984), Chaudhuri (1987, 1992), for which ϕ_i reduces to specific forms; their RR devices are also clearly specified but left unprescribed by Arnab (1995) to the extent that the model cited above remains tenable. Arnab (1990) offers arguments to ensure that E_R commutes with E_P, the operator for expectation with respect to a design p. Writing $d = (s,r), \underline{r} = (r_1, ..., r_i, ..., r_N)$ for the RR survey data at hand with r_i omitted in d for $i \in s$, he considers estimators for Y of the forms

$$e_u = e_u(d)$$

such that

$$E_p E_R(e_u) = E_R E_p(e_u) = Y \ \forall \ \underline{Y}, \ \underline{Y} = (y_1, ..., y_i, ..., y_N), \ e_1 = e_1(d) = b_s + \sum_{i \in s} b_{si} r_i,$$

b_s free of r, Y, b_{si}'s free of r, Y such that

$$\sum_s b_s p(s) = 0, \quad \sum_{s \ni i} b_{si} p(s) = 1 \ \forall i$$

and $e_2 = \sum_{i \in s} b_{si} r_i$, such that b_{si} is free of r, Y and $\sum_{s \ni i} b_{si} p(s) = 1 \ \forall i$. Each e_u, e_1, e_2 is unbiased, rather pR-unbiased for Y. Arnab (1995a) further postulates a superpopulation model M for which on writing E_m, V_m, C_m as expectation, variance, and covariance operators with respect to the probability distributions postulated for Y, he denotes

$$E_m(y_i) = \theta_i, \quad V_m(y_i) = \sigma_i^2$$

and

$$C_m(y_i, y_j) = \rho \sigma_i \sigma_j (i \neq j), \quad \sigma_i > 0$$

and

$$\left(-\frac{1}{N-1} \leq \rho \leq +1 \right).$$

In particular, he denotes M by M_0 when $\theta_i = \mu$ (unknown), $\sigma_i = \sigma$ (unknown), by M_1 when $\theta_i = \beta x_i$ (β unknown), $\sigma_i = \sigma x_i$ and M_2 when $\rho = 0$. He then derived the following results as generalizations over those of Godambe (1980), Chaudhuri and Adhikary (1981), and Chaudhuri (1987).

Theorem 9.1

Under M_0, it follows that

$$E_m E_{p_n} E_R (e_1 - Y)^2 \geq \sum_1^N \left[\frac{\left\{ E_m(\phi_i) + (1-\rho)\sigma_i^2 \right\}}{\pi_i} - (1-\rho)\sigma_i^2 \right] \quad \text{if } \rho > 0.$$

Theorem 9.2

Under M_0, if $\rho > 0$,

$$E_m E_{p_n} E_R (e_1 - Y)^2 \geq E_m E_{p_o} E_R (e_0 - Y)^2 = \frac{\left(\sum \sigma_i^* \right)^2}{n} - (1-\rho) \sum \sigma_i^2$$

writing $e_0 = p_0$ as a p_n for which π_i equals

$$\pi_i^* = \frac{n\sigma_i^*}{\sum_1^N \sigma_i^*} \tag{9.1}$$

and

$$\sum_{i \in s} \left(\frac{\theta_i}{\pi_i} \right) \text{ equals } \sum_1^N \theta_i \quad \forall s \text{ with } p(s) > 0; \tag{9.2}$$

here

$$\left(\sigma_i^* \right)^2 = E_m \left(\phi_i \right) + \left(1 - \rho \right) \sigma_i^2 \quad \text{and} \quad \left(\sum_1^N \sigma_i^* \right)^2$$

is the minimum value of $\sum_1^N (\sigma_i^*)^2 / \pi_i$ when π_i equals π_i^* above. To construct such a p_0 is of course extremely difficult.

A special case of p_0 is p_0^1 for which $\pi_i = (n/N) \; \forall \; i$ satisfies Equations 9.1 and 9.2 above. In fact, $E_m(\phi_i)$ is a constant for every i in U for the situations under M_0 covered by the results of Eriksson (1973), Godambe (1980), Chaudhuri and Adhikary (1981), Adhikary et al. (1984), and Chaudhuri (1987, 1992) as noted by Arnab (1995a). Next, writing $\bar{r} = (1/n)\Sigma_{i \in s} r_i$, he gets

---

Theorem 9.3

Under M_0,

$$E_m E_{p_n} E_R \left(e_2 - Y \right)^2 \geq E_m E_{p_0^1} E_R \left(N\bar{r} - Y \right)^2 = \frac{N \left[\phi_0 + \left(1 - \rho \right) \left(1 - \dfrac{n}{N} \right) \sigma^2 \right]}{n}$$

provided $E_m (\phi_i)$ is a constant $\phi_0 \; \forall \; i \in U$.

Note: Arnab (1995a) clarified that in Theorem 9.3, positivity is not required for ρ.

Remark

If M_0 in Theorem 9.2 is changed into M_1 with π_i as $\pi_i^* = nx_i/X$, $E_m(\phi_i) = Kx_i^2$ (K_{a+ve} constant) p_n changed into p_0^* for which π_i is replaced by π_i^*, then follows

Theorem 9.4

Under M_1

$$E_m E_{p_n} E_R \left(e_1 - Y\right)^2 \geq E_m E_{p_0'} E_R \left(e_0 - Y\right)^2$$

$$= \left[K + (1-\rho)\sigma^2\right]\frac{X^2}{n} - (1-\rho)\sum_1^N x_i^2\right] \quad \forall \, \rho \in \left[-\frac{1}{N-1}, +1\right]$$

Theorem 9.5

Finally, under M_2, he gets

$$E_m E_{p_n} E_R \left(e_u - Y\right)^2 \geq \sum_1^N \left[\left\{\sigma_i^2 + E_m \phi_i\right\}/\pi_i\right] - \sum_1^N \sigma_i^2$$

Compare Chaudhuri and Adhikary (1981) and Chaudhuri (1987).

Arnab (1996) provides a unified theory for the estimation of the proportion in a community bearing a stigmatizing attribute covering any sampling design, wide classes of unbiased estimators and broad classes of RR devices, offering unbiased variance estimators as well.

As usual A is a stigmatizing attribute, A^c its complement and a person i in $U = (1, ..., i, ..., N)$ bears a value y_i which is 1 if i bears A and is 0 if i bears A^c and the target parameter to estimate is $Y = \sum_1^N y_i$. Also p is a design with $p(s)$ as the probability to choose a sample s, $\pi_i = \sum_{s \ni i} p(s)$ and $\pi_{ij} = \sum_{s \ni i, j} p(s)$ are the inclusion probabilities for i and the pair (i, j), $i \neq j$. Also $\pi_i > 0 \; \forall \, i$ and $\pi_{ij} > 0 \; \forall \, i \neq j$ as is required to be assumed for unbiased estimation of Y and of the variance of any such estimator of Y. Arnab (1996) restricted his study to homogeneous linear unbiased estimators for Y of the form

$$t_b = \sum_{i \in s} b_{si} y_i = t_b\left(s, \underline{Y}\right)$$

with b_{si}'s as constants free of $\underline{Y} = (y_1, ..., y_i, ..., y_N)$ and of $\underline{r} = (r_1, ..., r_i, ..., r_N)$ with r_i's as independent variables with $E_R(r_i) = y_i$, $V_R(r_i) = \phi_i$ and $C_R(r_i, r_j) = 0 \; \forall \, i \neq j$, writing E_R, V_R, C_R as operators for expectation, variance, and covariance with respect to suitable RR devices yielding these r_i's for $i \in U$. Further,

$$\sum_{s \ni i} b_{si} p(s) = 1 \quad \forall \, i \in U,$$

E_p, V_p are operators for taking expectation, variance with respect to the design p. Also

$$\phi = \sum_1^N \phi_i, \quad E_p(t_b) = Y, \quad V_p(t_b) = \sum y_i^2 \left(\alpha_i - \frac{1}{N^2} \right) + \sum_{i \neq} \sum_j y_i y_j \left(\alpha_{ij} - \frac{1}{N^2} \right)$$

with

$$\alpha_i = \sum_{s \ni i} b_{si}^2 p(s), \quad \alpha_{ij} = \sum_{s \ni i,j} b_{si} b_{sj} p(s) \quad \text{and} \quad \hat{V}_p(t_b) = \sum_{i \in s} C_{si} y_i^2 + \sum_{i \neq} \sum_{j \in s} C_{sij} y_i y_j$$

with C_{si}, C_{sij} as constants free of $\underline{Y}$, subject to

$$\sum_{s \ni i} C_{si} p(s) = \alpha_i - \frac{1}{N^2} \quad \text{and} \quad \sum_{s \ni i,j} C_{sij}(p(s)) = \alpha_{ij} - \frac{1}{N^2}$$

so that $\hat{V}_p(t_b)$ is an unbiased estimator for $V_p(t_b)$. Since t_b, $\hat{V}_p(t_b)$ are not usable as y_i's are not available his recommended estimator for Y is $e_b = \sum_{i \in s} b_{si} r_i$ whose variance is

$$V(e_b) = V_p(t_b) + E_p \sum_{i \in s} b_{si}^2 \phi_i$$

of which an unbiased estimator is

$$\hat{V}(e_b) = \sum_{i \in s} C_{si} r_i^2 + \sum_{i \neq j} \sum_{\in s} C_{sij} r_i r_j + \frac{1}{N^2} \hat{\phi}$$

where $\hat{\phi}$ is a suitably chosen unbiased estimator for $\phi = \sum_1^N \phi_i$, for example, as $\hat{\phi} = \sum_{i \in s} \hat{\phi}_i / \pi_i$ with $\hat{\phi}_i$ as a suitably chosen unbiased estimator for ϕ_i, $i \in s$. Arnab (1996) was motivated to publish this work to improve upon the studies conducted by Franklin (1989b) and Singh and Singh (1993), who restricted their procedures to SRSWR alone.

Our principal comments on Arnab's (1996) work are that (1) he did not utilize the fact $y_i^2 = y_i$, which renders $V_p(t_b)$ as

$$V_p'(t_b) = \sum_1^N y_i \left(\alpha_i - \frac{1}{N^2} \right) + \sum_{i \neq j} \sum_{\in U} y_i y_j \left(\alpha_{ij} - \frac{1}{N^2} \right),$$

which immediately yields the unbiased estimator

$$\tilde{V}_p(t_b) = \sum_{i \in s} C_{si} y_i + \sum_{i \neq} \sum_j C_{sij} y_i y_j$$

leading to $V(e_b)$ revised as

$$V'(e_b) = V'_p(t_b) + \sum \phi_i \left(\alpha_i - \frac{1}{N^2} \right),$$

which admits an unbiased estimator

$$\tilde{V}'(e_b) = \sum_{i \in s} C_{si} r_i + \sum_{i \neq} \sum_j C_{sij} r_i r_j + \frac{1}{N^2} \hat{\phi},$$

and in addition (2) he did not attempt at establishing any optimality results in this situation with $y_i = 1/0$ only with or without postulating any super-population modeling for $\underline{Y} = (y_1, \ldots, y_i, \ldots, y_N)$ with this binary restriction. Arnab (1995b, p. 4) presents the following results that are not true for open surveys but are true for RR surveys: (1) admissibility of e_{HH} in the class of linear pR unbiased estimators under PPSWR sampling scheme; (2) admissibility of sampling strategy based on sample mean of $r_i(k)$'s in SRSWR scheme in the class of strategies $H^* = (e, p), e \in C, p \in p_1$ and (3) for a noncensus UCSD, the Horvitz and Thompson (1952) estimator is not the minimum variance unbiased estimator (MVUE) for Y. The notations used are as given by Arnab (1996).

We fail to appreciate the author's presentation of his claim of a crucial result (2.2) ibid based on Cauchy–Schwarz inequality and that of $E_p\{n_i(s)\} = np_i$ for PPSWR, $n_i(s)$ = frequency of ith unit in s is obviously untenable leading to his overall proofs being of questionable values.

In fact,

$$E_p \{n_i(s)\} = \sum_{i=1}^{N} \left[1 - (1 - p_i)^n \right] \neq np_i.$$

We fail to accept his Theorems 9.1 and 9.2 as described earlier.

Referring to Arnab (1998, 2000) it seems useful to continue with his contributions to RR techniques with the notations already familiar to the readers. Arnab (1998) observes for $E_R(r_i) = y_i$, $V_R(r_i) = \phi_i > 0$, $C_R(r_i, r_j) = 0 \; \forall \; i \neq j$, $E_m(y_i) = \beta x_i$, $V_m(y_i) = \sigma^2 x_i^2$, $C_m(y_i, y_j) = 0 \; \forall \; i \neq j$ that

$$V(e_b) = E_p E_R(e_b - Y)^2 = E_p \left(\sum_{i \in s} b_{si}^2 \phi_i \right) + V_p \left(\sum_{i \in s} b_{si} y_i \right) \geq \sum_{i=1}^{N} \frac{\phi_i}{\pi_i} + V_p(t_b).$$

Also he recalls that for $t_H = \sum_{i \in s} (y_i/\pi_i)$, the Horvitz and Thompson's (1952) estimator for Y, in case it is based on a design p_{nx} for which every sample has a fixed effective size n and $\pi_i = nx_i/X$, that is, it is a π PS or IPPS design, $E_m V_p(t_b) \geq E_m V_{pnx}(t_H)$. Then, $e_H = \sum_{i \in d} (r_i/\pi_i)$ based on p_{nx} gives an optimal RR strategy because $E_m V(e_b) \geq E_m E_{pnx} E_R (e_H - Y)^2$ provided $E_m(\phi_i) \alpha x_i^2$. Next it is instructive to pay heed to what Arnab (1998) does to achieve $E_m(\phi_i) \alpha x_i^2 \ \forall i \in U$ which is his condition for the optimality of an RR strategy.

First, he adopts Sinha and Hedayat's (1991) dictum, without acknowledging it however, to presume the possibility that a finite number of L quantities $K_1, ..., K_L$ may be found by a person i so that the value y_i is covered by the numbers $K_j x_i$ for $j = 1, ..., L$, when x_i is the same as the number as in the model he postulates. With this presumption he modifies Eriksson's (1973) RR technique by demanding a responding person i sampled to give out the true value y_i with a probability c ($0 < c < 1$) or to give out the value $K_j x_i$ with probability q_j ($0 < q_j < 1$) so that $\Sigma_{j=1}^{L} q_j = 1 - c$, this being independent for every i in U. Then, taking z_i as the RR from i, if sampled, and

$$r_i = \frac{\left[z_i - x_i \sum_{j=1}^{L} q_j K_j \right]}{c}, \quad i \in U$$

it follows that $E_R(r_i) = y_i$ and $E_m V_R(r_i) = E_m(\phi_i) = \delta x_i^2$, where

$$\delta = \frac{1}{c^2} \left[c(1-c)(\sigma^2 + \beta^2) - 2\beta c \left(\sum_1^L K_j q_j \right) - \left(\sum_1^L K_j^2 q_j \right) + \left(\sum K_j q_j \right)^2 \right].$$

He also showed that the same optimal RR strategy can be achieved by modifying Chaudhuri's (1987) RR technique for quantitative variables with a very little effort even without resorting to Sinha and Hedayat's (1991) approach. He further showed that in case x_i is taken as unity for every i in U, then on implementing these modifications on either Eriksson's (1973) or Chaudhuri's (1987) RR device the sample mean of the corresponding r_i's with $E_R(r_i) = y_i$ in combination with SRSWOR provides an optimal RR strategy under his postulated model. He further achieves the same strategy as optimal for his modified Eriksson (1973) or Chaudhuri (1987) RR device if for $\underline{Y} = (y_1, ..., y_i, ..., y_N)$ he postulates the random permutation model (RPM) for which each of those $N!$ possible permutations $(y_{i1}, ..., y_{ij}, ..., y_{iN})$ corresponding to $(i_1, ..., i_j, ..., i_N)$ occurs with an equal probability $1/N!$ because under this RPM, one obviously has $E_m(y_i) = \bar{Y} = 1/N \Sigma_1^N y_i$ and $V_m(y_i) = 1/N \Sigma_1^N (y_i - \bar{Y})^2$, $C_m(y_i, y_j) = -1/N(N-1)$ so that $E_m(\phi_i)$ is a constant.

These results are further generalized by Arnab (2001) himself using the RR devices modified as above on Eriksson (1973) and Chaudhuri (1987) but by postulating the four super-population models on $\underline{Y} = (y_1, ..., y_i, ..., y_N)$ namely M_1, M_2, M_3, M_4 and using certain results on DR data given by Chaudhuri and Stenger (1992).

Let model M: $E_m(y_i) = \mu_i$, $V_m(y_i) = v_{ii}(>0)$

$$C_m\left(y_i, y_j\right) = v_{ij}, \quad i \neq j, \quad i, j = 1, \dots, N$$

For M_1 : $v_{ij} = \rho\sqrt{v_{ii}\, v_{jj}}$,

For M_2 : $\mu_i = \beta x_i$, $v_{ii} = \sigma^2 x_i^2$, $v_{ij} = \sigma^2 \rho x_i x_j$
For M_3 : $\mu_i = \mu$, $v_{ii} = \sigma^2$, $v_{ij} = \rho\sigma^2$, $i \neq j$
For M_4 : y_i's are independent; μ_i, β, $\sigma(>0)$, ρ unknown but $(-1 < \rho < +1)$. An estimator $t = t(s, \underline{R})$ based on a sample s chosen according to a design p for $Y = \sum_1^N y_i$ gives the strategy $h = (p, t)$ and H is a class of strategies. For an RR survey, a strategy (p_0, t_0) in H is optimal if

$$E_m E_{p_0} E_R \left(t_0\left(s, \underline{R}\right) - Y\right)^2 \le E_m E_p E_R \left(t\left(s, \underline{R}\right) - Y\right)^2$$

for a (p_0, t_0) and (p, t) anyone in H; here of course $\underline{R} = (r_1, \dots, r_i, \dots, r_N)$.

To work out $e_0(s, \underline{R})$ he restricts to $e(s, \underline{R}) = a_s + \sum_{i \in s} b_{si} r_i$ subject to $E_p(a_s) = 0$, $\sum_{s \ni i} b_{si} p(s) = 1 \; \forall\, i \in U$ with a_s, b_{si} free of $\underline{R} = (r_1, \dots, r_N)$. Let $V = (v_{ij})$ be an $N \times N$ matrix of the v_{ij}'s, $V_s =$ a submatrix of V for i, j in s, V_s^{-1}, the inverse of V_s with v_s^{ij} as the entries of V_s^{-1}, $\delta_{ij} = \sum_{s \ni ij} v_s^{ij} p(s)$, $\Delta = ((\delta_{ij}))$, $\Delta^{-1} = ((\delta^{ij}))$

$$\underline{\lambda} = \left(\lambda_1, \dots, \lambda_N\right) = \Delta^{-1} I$$

$I = N \times 1$ column matrix of unities, $\underline{\lambda}s = n \times 1$ subvector of $\underline{X}$ with only the entries λ_i, $i \in s$, n is the constant size of s. Then he quotes the following results from Chaudhuri and Stenger (1992) concerning DR-based estimators $e = e(s, \underline{Y}) = a_s + \sum_{i \in s} b_{si} y_i$ and designs p and p_n's the latter restricting to samples each with n distinct units.

Result 1

$$E_m E_p \left[e(s, \underline{Y}) - Y\right]^2 \ge \sum\sum \delta^{ij} - \sum\sum v_{ij} = E_m E_p \left[e^*(s, \underline{Y}) - Y\right]^2,$$

$$e^* = e^*(s, \underline{Y}) = \sum b_{si}^*(y_i - \mu_i) + \sum_1^N \mu_i,$$

$$\underline{b}_s^* = V_s^{-1} \underline{\lambda} s \quad \text{with } b_{si}^* \text{ as its elements.}$$

Let $e(s, \underline{R}) = a_s + \sum_{i \in s} b_{si} r_i$

$$E_p E_R \left[e(s, \underline{R}) - Y \right]^2 = E_p \left\{ V_R \left[e(s, R) \right] \right\} + E_p \left\{ E_R \left[e(s, \underline{R}) \right] - Y \right\}^2$$

$$= \sum_i \phi_i \left(\sum_{s \ni i} b_{si}^2 p(s) \right) + E_p \left[e(s, \underline{Y}) - Y \right]^2$$

and

$$E_m E_p E_R \left[e(s, \underline{R}) - Y \right]^2 = \sum_i E_m (\phi_i) \left[\sum_{s \ni i} b_{si}^2 p(s) \right] + E_m E_p \left[e(s, \underline{Y}) - Y \right]^2.$$

Theorem 9.6

For a given $p \in p_n$,

$$E_m E_p E_R \left[e(s, \underline{R}) - Y \right]^2 \geq \sum_i \sum_j \delta^{ij} - \sum_i \sum_j v_{ij} + \sum_i E_m (\phi_i / \pi_i)$$

Writing E_{m_1} for E_m using model M_1 and utilizing Mukerjee and Sengupta's (1989) work as also reported by Chaudhuri and Stenger (1992), Arnab (2000) states:

Result 2

$$E_{m_1} E_p \left[e(s, \underline{Y}) - Y \right]^2 \geq \sum_i \sum_j \delta^{ij} - \sum_i \sum_j v_{ij} \geq (1 - \rho) \left[\frac{\left(\sum \sqrt{v_{ii}} \right)^2}{n - \sum v_{ii}} \right]$$

$$= E_{m_1} E_{p0} \left[e_0 (s, \underline{Y}) - Y \right]^2$$

where

$$e_0 (s, \underline{Y}) = \sum_{i \in s} \frac{(y_i - \mu_i)}{\pi_{i0}} + \sum_i \mu_i$$

and p_0 is a sampling design with inclusion probability $\pi_{i0} = n \sqrt{v_{ii}} / \sum_i \sqrt{v_{ii}}$.

This leads to

$$
E_{m_1} E_p E_R \left[e(s, \underline{R}) - Y \right]^2 \geq (1 - \rho) \left[\frac{\left(\sum_i \sqrt{v_{ii}} \right)^2}{n - \sum v_{ii}} \right] + \sum E_{m_1} \frac{(\phi_i)}{\pi_i}.
$$

The term $\sum E_{m_1}(\phi_i)/\pi_i$ is minimized for $\pi_i = n \sqrt{E_{m_1}(\phi_i)} / \sum_i \sqrt{E_{m_1}(\phi_i)}$ particularly

$$
\sum_1^N \pi_i = n.
$$

He also notes

$$
E_{m_1} E_p E_R \left[e(s, \underline{R}) - Y \right]^2 \geq (1 - \rho) \left[\frac{\left(\sum_i \sqrt{v_{ii}} \right)^2}{n - \sum_1^N v_{ii}} \right] + \frac{\left[\sum_i \sqrt{E_{m_1}(\phi_i)} \right]^2}{n} \tag{9.3}
$$

The lower bound of Theorem 9.6 is attained by a sampling strategy (e_0, p_0^*) with

$$
e_0 = e_0(s, \underline{Y}) = \sum_{i \in s} \frac{(y_i - \mu_i)}{\pi_{i0}} + \sum_1^N \mu_i
$$

and p_0^* is a design with inclusion probability

$$
\pi_{i0}^* = \frac{n \sqrt{v_{ii}}}{\sum_i \sqrt{v_{ii}}} = \frac{n \sqrt{E_{m_1}(\phi_i)}}{\sum_i \sqrt{E_{m_1}(\phi_i)}} \tag{9.4}
$$

Hence, follows:

Theorem 9.7

The strategy $h_0 = (e_0, p_0^*)$ is optimum in the class $H = \{(e, p)\}$ for $e = a_s + \sum_{i \in s} b_{si} r_i$ and p as a p_n.

Employing M_2 and writing E_{m_2} for E_m when M is M_2, one gets

$$E_{m_2} E_p E_R \left[e(s, \underline{R}) - Y \right]^2 \geq (1 - \rho) \sigma^2 \left[\frac{X^2}{n} - \sum_1^N x_i^2 \right] + \frac{\left[\sum_i \sqrt{E_m(\phi_i)} \right]^2}{n} \quad (9.5)$$

$$= E_{m_2} E_{\tilde{p}} E_R \left[\sum_{i \in s} \frac{r_i}{\pi_i} - Y \right]^2$$

writing $\tilde{p}$ for a design p for which $\pi_i = n x_i / X = n \sqrt{E_{m_2}(\phi_i)} / \sum_1^N \sqrt{E_{m_2}(\phi_i)}$ and $E_{\tilde{p}}$ for E_p when p is $\tilde{p}$.

To get a strategy $h = (e, p)$ for which the lower bound on the right-hand side (RHS) in Equation 9.5 may be attained, an RR device has to be implemented for which

$$E_{m_2}(\phi_i) \propto x_i^2.$$

An RR achieving this will be denoted as R_0. Hence,

Theorem 9.8

$$E_{m_2} E_{p_n} E_{R_0} \left[\sum_{i \in s} b_{si} r_i - Y \right]^2 \geq E_{m_2} E_{\tilde{p}} E_{R_0} \left[\sum_{i \in s} \frac{r_i}{\pi_i} - Y \right]^2$$

Arnab (1998) explained how an R_0 may be devised by initiating his study with Eriksson (1973) and Chaudhuri (1987).

Let R_0^* be an RRT for which $E_m \phi_i = \phi_3$, a constant, independent of i and E_{m_3} be E_m and V_{m_3} be V_m when M is M_3.

Then, writing $\bar{p}$ for an SRSWOR design Arnab (2001) derives:

$$E_{m_3} E_p E_{R_0^*} \left[e(s, \underline{R}) - Y \right]^2 \geq (1 - \rho) \sigma^2 \left[\frac{N^2}{n} - N \right] + \frac{\left[N E_{m_3}(\phi_0) \right]^2}{n}$$

$$= E_{m_3} E_{\bar{p}} E_{R_0^*} \left[N \bar{r}_s - Y \right]^2$$

bestowing an optimality result on the strategy

$$\left(\bar{r}_s = \frac{1}{n} \sum_{i \in s} r_i, \bar{p} \right).$$

For the model M_4, Godambe and Joshi (1965) brought forth the result

$$E_{m_4} E_{Pn} \left[e(s, \underline{\tilde{R}}) - Y \right]^2 \geq \sum_i v_{ii} \left(\frac{1}{\pi_i} - 1 \right) = E_{m_4} E_{\tilde{p}} \left[\sum_{i \in s} \frac{r_i}{\pi_i} - Y \right]^2$$

writing E_{m_4} for E_m when M is M_4 and $e(s, \underline{R})$ as an unbiased estimator for $R = \sum_1^N r_i$.

Recognizing that

$$E_{m_4} E_R \left(r_i \right) = \mu_i \quad \text{and} \quad E_{m_4} E_R \left[r_i - Y \right]^2 = E_{m_4} \left[V_R \left(r_i \right) \right] + V_{m_4} \left[E_R \left(r_i \right) \right]$$
$$= E_{m_4} \left(\phi_i \right) + V_{m_4} \left(y_i \right),$$

Arnab (2000) derives

$$E_{m_4} E_p E_R \left[\tilde{e}(s, \underline{R}) - Y \right]^2 = E_{m_4} E_R E_p \left[\tilde{e}(s, \underline{R}) - \sum_1^N r_i \right]^2 + E_{m_4} E_R \left[\sum_1^N r_i - Y \right]^2$$
$$= E_{m_4} E_R E_p \left[\tilde{e}(s, \underline{R}) - \sum_1^N r_i \right]^2 + E_{m_4} \left[V_R \left(\sum_1^N r_i \right) \right].$$

Writing π_{i0} for π_i for which

$$\pi_i = \frac{n \sqrt{v_{ii} + E_{m_4} \left(\phi_i \right)}}{\sum_i \sqrt{v_{ii} + E_{m_4} \left(\phi_i \right)}},$$

Arnab (2001) shows the optimality of the strategy $\left(\tilde{e}(s, \underline{R}), \tilde{p}_j \right)$ among $\left(\tilde{e}(s, \underline{R}), p_n \right)$ in the sense that

$$E_{m_4} E_{p_n} E_R \left[\tilde{e}(s, \underline{R}) - Y \right]^2 \geq E_{m_4} E_{\tilde{p}0} E_R \left[\sum_{i \in s} \frac{r_i}{\pi_i} - Y \right]^2.$$

This $\tilde{p}_0$ is usable in practice only if $E_{m_4} (\phi_i) \propto x_i^2, i \in U$. This situation is realized on modifying Eriksson's (1973) RRT and adjusting the permissible values of y_i's as discussed by Arnab (1998). So, this final result of Arnab (2001) is subject to the same criticism as noted in connection with the corresponding result of Arnab (1998).

One deficiency in the optimal sampling strategies recounted by Arnab is that no clue is given about how to estimate suitable measures of errors of the respective estimators.

Arnab (2004) considered virtually the same optional RR procedure as treated by Chaudhuri and Mukerjee (1985, 1988) for SRSWR and of Chaudhuri and Saha (2005). We shall discuss his results in connection with those by other contributors to optional randomized response (ORR).

Nayak, Nayak and Adeshiyan, Christofides, and Quatember

Nayak's (1994) work is highly important in the RR literature because although he restricts his study to selection of possible respondents by SRSWR method alone and to the estimation of the proportion θ bearing a stigmatizing characteristic A, in a dichotomous population, he only seeks a suitable procedure of estimation on applying the broad-based twin criterion that (1) a respondent's privacy must be protected rationally and (2) subject to this desideratum an efficient estimator with a controlled measure of accuracy must be sought. Allowing a mere two-fold response structure as "Yes" or "No" about bearing A or its complement A^c respectively and taking "Yes" or Y as 1 and "No" or N as 0 respectively he denotes the $Prob(Y \mid A)$ by "a" and $Prob(Y \mid A^c)$ by "b." Since SRSWR alone is permitted, the prior probability that any person chosen on any draw from the population of persons in the community is $P(A) = \theta$. So, by Bayes' theorem he gets the posterior probability that a person bears the sensitive characteristic A given that a truthful response "Yes" or "No" is given on employing a specific RR device as

$$P(A \mid Y) = \frac{a\theta}{a\theta + (1 - b)(1 - \theta)}$$

and

$$P(A \mid N) = \frac{(1 - a)\theta}{(1 - a)\theta + b(1 - \theta)}.$$

These are covered by Warner's (1965) pioneering RR model and Simmons' model as described by Greenberg et al. (1969) with the probability that a person sampled may bear an innocuous feature B unrelated to A be known and truthfully announce whether A or A^c or B or B^c is borne by the person.

Nayak (1994) recognizes a and b above as the "Design parameters" of an RR device. His motive is to develop a unified theory for RR surveys in the set up noted above. Writing p as the observed proportion of the persons saying "Yes" in an SRSWR in n draws he claims

$$\hat{\theta} = \frac{(p + b - 1)}{(a + b - 1)}$$

as the minimum variance unbiased estimator (MVUE) for θ with a variance

$$V(\hat{\theta}) = \frac{\lambda(1 - \lambda)}{\left[n(a + b - 1)^2 \right]},$$

$$\lambda = a\theta + (1 - b)(1 - \theta) = (1 - b) + (a + b - 1)\theta,$$

so long as a and b are elements in the "design-space" denoted as

$$D = \left\{ (a,b) : 0 \le a,b \le 1,\ 1 < (a + b) < 2 \right\}$$

Nayak (1994) characterized certain RR procedures in terms of their equivalence and lack of equivalence. Writing d_1 and d_2 as two RR designs with $P_{d_1}(A|Y)$, $P_{d_2}(A|Y)$, $P_{d_1}(A|N)$ $P_{d_2}(A|N)$, and $V_{d_1}(\hat{\theta})$, $V_{d_2}(\hat{\theta})$ defined "Best" designs d in D, admissible designs in D. Flinger et al. (1977) is a relevant reference for these works. Defining $u = a/(1 - b)$ and $v = b/(1 - a)$, Nayak (1994) noted an equivalent "Design-space" $\hat{D} = \{(u,v)|1 < u, v < \infty, (u,v) \ne (\infty,\infty)\}$. He also noted $V(\hat{\theta})$ as a decreasing function of both u and v and equivalently of both a and b. He also noted the crucial fact that both efficiency and "respondent's protection" are best achieved if $v = +\infty$, equivalently $a = 1$ and more noteworthily $D_0 = \{(a,b)|a = 1, 0 < b < 1\}$ as the class of all admissible designs. He also observes, contrary to a popular belief that efficiency and privacy protection need not move in opposite directions.

As a rider he observed that Warner's (1965) model is inadmissible and so also is Simmons's model with a known $\beta = Prob$ [a person bears B] unless $\beta = 1$.

As discussed by Chaudhuri and Mukerjee (1988), a requirement of an optimal design as per the theories developed by Lanke (1975b), Loynes (1976), and Leysieffer and Warner (1976) is that $a = 1$ and in addition $P(A|N) = 1$.

Nayak (1994) measures privacy by the probabilities of classifying a respondent bearing A for both "Yes" and "No" responses but the above mentioned authors use a single quantity such as $P(A|Y)$ or $\max[P(A|Y), P(A|N)]$ to measure the "jeopardy" carried by an RR device. But the status of $P(A|Y)$ described by Lanke (1975) "The risk of suspicion" is quite high in this context. Flinger et al. (1977) call $[1 - \max\{P(A|Y), P(A|N)\}]/(1 - \theta)$ a "measure of primary protection."

Further, theoretically this particular work by Nayak (1994) also deserves attention from those interested in RR techniques.

More audaciously provocative approach of RRT to estimate a proportion θ of people bearing the sensitive attribute A in a dichotomous population from binary or polychotomous responses from people selected with unequal probabilities has been dealt with in a unified way by Nayak and Adeshiyan (2009).

They start by referring to Warner's (1965) and Simmons' unrelated question models with known probability of a person's bearing an innocuous feature B unrelated to A and note that if for the two RR procedures, the latter described by Greenberg et al. (1969), the probability that a person is asked to report if the person's feature is A be taken the same amount $p(0 < p < 1, p \neq (1/2))$, then although the Warner's RRT is worse between the two in terms of variance of the unbiased estimator of θ from a sample taken by SRSWR in the same number of draws as n for both, then the latter is superior as it affords a greater protection of privacy. They recommend comparing two RR procedures characterized by various parameters prescribing the respective RR designs in terms of the variances of suitably unbiased estimators of θ keeping the respective measures of protection of privacy at par.

Earlier Lanke (1975b), Loynes (1976), Leysieffer and Warner (1976), and Fligner et al. (1977) adopted their measures of privacy protection and Nayak (1994) himself prescribed different ones with a Bayesian concept of posterior probability of a person bearing A with an unknown prior probability combined with an RR produced by a person sampled and addressed. Each of these stalwarts restricted himself to the selection of respondents by SRSWR scheme alone.

Although the population addressed is dichotomous, RRs used to estimate θ are permitted to be binary or polychotomous. Nayak and Adeshiyan (2009) painstakingly developed a unified theory to show that given an estimator based on polychotomous RRs equivalent in terms of equally efficient estimators based on alternative binary RR-based procedures are available with measures of protected privacy being kept on par.

Nayak and Adeshiyan (2009) also covered RRs emanated from samples chosen with unequal selection probabilities and sensitive variables both qualitative and quantitative in nature.

Unfortunately, both Nayak (1994) and Nayak and Adeshiyan (2009) have so far remained silent about how to modify the rules for deriving the posterior probabilities needed in prescribing the measures of protection when covering qualitative characteristics with SRSWR sampling replaced by any other sampling scheme and when covering stigmatizing characteristics that are quantitative. In Chapter 7, we described our procedures to include general sampling schemes and the qualitative characteristics. But we are yet to see any solution for quantitative variables that are sensitive no matter how a sample of respondents is selected. To us it still seems a tall order. But the energetically spelled out coverages by Nayak (1994) and Nayak and Adeshiyan (2009) of their unified theories still continue to remain deficient in this respect. They refuted the claim by Christofides (2003) that his proposed alternative RR device permitting polychotomous responses may fare better than the Warner's with a wider range of choice of RRT parameters by showing errors in the latter's numerical calculations. Christofides (2010) took a strong exception by citing counterexamples in the opposite direction. Nayak and Adeshiyan (2009) however, contended that in order to keep the "protected privacy" under check multichotomous RR devices are redundant in achieving desired efficiency level to contest binary RR-based procedures. It is instructive to place on record that Chaudhuri's (2004) version of Christofides's (2003) RRT when the latter's SRSWR based treatment is revised offering unequal probability sampling alternatives, though cited by Nayak and Adeshiyan (2009) is not commented upon at all by them in respect of privacy protection-cum efficiency level attainment. In Chapter 7 of this book a necessary explanation has been provided to attract due attention.

Quatember (2009) in the meanwhile has come up with vigorous opposition to the SRSWR based RRT of Christofides by highlighting its numerical shortcomings. To this no response seemed warranted except noting his silence on the revised RRT of Chaudhuri (2004) extending it to general sampling designs.

Guerriero and Sandri

Guerriero and Sandri (2007) are also two well-known votaries that any two RR procedures should be compared only under the twin criteria of (1) efficient estimation in terms of suitable estimators of sensitive features from RRs and (2) the level of protection afforded by them of the privacies of the respondents. Their sole target parameter demanding estimation is the finite population proportion of people namely θ in a community bearing a sensitive characteristic A. Their goal is also to treat a binary RR for a dichotomous population. They define $P(R|A)$ with R as "Yes"/"No" response as "design probabilities" as usual. They take privacy protection (PP) measures following the traditional approach as in Lanke (1975b, 1976) and Leysieffer and Warner (1976) and consider estimators $\hat{\theta}$ for θ based on RRs from SRSWRs in n draws. They examine relative performances of RR procedures with respect to efficiency of an estimator alone and again separately in terms of efficiency-cum-privacy protection. Algebra is too cumbrous to bear even partial reproduction here.

Padmawar and Vijayan

Padmawar and Vijayan (2000) term our RR approach covering a finite population $U = (1, \ldots, i, \ldots, N)$ as described in Chapter 7, a traditional approach and present an opposing one briefly summarized below.

They regard y_i as the value of a study variable y for the unit i in U, supposed to be a stigmatizing one, described as a "signal" and get an RR as a mixture of a signal and a "noise." This noise value is x_i for i generated from a known distribution with mean μ and variance σ^2 both known as constant for every i in U. The RR from i is

$$z_i = y_i + x_i,$$

for i in a sample s chosen according to a suitable design p. Here there is no restriction on the nature of x_i's vis-à-vis the y_i values. In the so-called traditional RR set up dealt with in Chapter 7 following Eriksson (1973), Godambe (1980), Chaudhuri et al. (1981), Chaudhuri (1987, 1992), Sengupta and Kundu (1989) and also explored by Arnab as reported in this chapter, the x_i's must have a range coinciding with that of y_i's, for example, especially, if $0 \le y_i \le 1 \, \forall i$, the same must hold for $x_i, i \in U$, raising a respondent's suspicion in a contrary case. But strikingly, Padmawar and Vijayan (2000) strongly rule out such a restriction altogether. Their claim is a respondent's total faith in the "privacy being protected doubtless" despite the RR as $z_i = y_i + x_i$ for every i in s. The aim is to derive an unbiased estimator for $Y = \sum_1^N y_i$ derived from the RR data as $(s, \hat{y}_i | i \in s)$ on developing $\hat{y}_i = z_i - \mu$ satisfying $E_R(\hat{y}_i) = y_i \, \forall i \in U$ persisting with the notations developed so far. They restrict to the class of DR based estimators for Y of the form

$$e = e(s, \underline{Y}) = \sum_{i \in s} b_{si} y_i$$

with b_{si}'s free of $\underline{Y} = (y_1, \ldots, y_i, \ldots y_N)$ and of $\underline{\hat{Y}} = (\hat{y}_1, \ldots, \hat{y}_i, \ldots \hat{y}_N)$ and subject to $\sum_s p(s) b_{si} = 1 \forall i$ and to the RR-data $s, \hat{y}_i (i \in s)$-based estimators for Y of the form

$$\hat{e} = \hat{e}(s, \underline{\hat{Y}}) = \sum_{i \in s} b_{si} \hat{y}_i$$

Their results worthy of attention are:

Theorem 9.9

$\hat{e} = \hat{e}(s, \underline{\hat{Y}})$ is unbiased for Y in the sense $E_p E_R(\hat{e}) = Y$.
  Writing $V_p(e)$, the variance of e based on DR-data $(s, y_i | i \in s)$ and $\hat{V}_p(e)$ as an unbiased estimator thereof;

Theorem 9.10

$$\hat{V}_p(e)\big|_{Y=\hat{Y}} + N\sigma^2 = \hat{V}(\hat{e})$$

is an unbiased estimator for $V(\hat{e})$ in the sense that

$$E(\hat{V}(\hat{e})) = V(\hat{e}) = V_p(e) + N\sigma^2$$

Further, taking $\underline{Z} = (z_1, ..., z_i, ..., z_N)$, $e(s, \underline{Z}) = \Sigma_{i \in s} b_{si} z_i$ another class of unbiased estimators for Y is $e(s, \underline{Z}) - N\mu = e_z$, say

$$V(e(s, \underline{Z})) = V_p(e(s, \underline{Y}) + \underline{\mu}) + \sigma^2 \sum_{i=1}^{N} E_p(b_{si}^2), \quad \underline{\mu} = \mu 1', \; 1' = (1, ..., 1).$$

Theorem 9.11

e is better than e_z if f

$$V_p(e) \le V_p(e_z).$$

Further $\hat{V}_p(e(s, \hat{Y})| \; \hat{\underline{Y}} + \underline{\mu}) - N\sigma^2$ is an unbiased estimator of $V(e_z)$.

Theorem 9.12

For a given p if $e(s, \underline{Y})$ is an admissible estimator for Y in a DR setup, then $e(s, \hat{Y})$ as derived from $e(s, \underline{Y})$ is also admissible.

Theorem 9.13

There does not exist a uniformly minimum variance linear unbiased estimator (UMVLUE) for Y in the RR setup.

Superpopulation model setup is also considered by Padmawar and Vijayan (2000) establishing a couple of optimality results in their RR setup. It is important for someone to take up a project to critically compare their work vis-à-vis those mentioned in Chapter 7 of this book and also with Arnab's approaches treated earlier in this chapter.

Works on RR by N.S. Mangat, Ravindra Singh, Sarjinder Singh, Sat Gupta, and Bhisham Gupta

The above-mentioned contributors to RR theory and practice except Sarjinder Singh restrict to "equal probability sampling" alone. Mangat and Singh (1991) consider estimating the sensitive proportion $\theta(0 < \theta < 1)$ from an SRSWOR in n draws rather than from an SRSWR. Such an exercise was earlier done by Kim and Flueck (1978) as was reported by Chaudhuri and Mukerjee (1988). We believe these activities may be explained on referring to Chaudhuri's (2001a) works. The crucial points are (1) an RR device yields data no matter how a person is chosen in a sample and (2) the probability that a person bears a sensitive attribute "*A*," no matter on which draw the person is selected, equals θ only in SRSWR. Comparison of various RR-based estimators for θ becomes difficult and inappropriate when RR data are gathered from unequal probability samples.

Mangat and Singh's (1990) work modifying the pioneering RRT by Warner (1965) has been mentioned in Chapter 3 of this book. Envisaging truthful reporting by a person in an SRSWR in *n* draws about bearing a stigmatizing attribute *A* by dint of a carefully devised RR technique they recommend adopting the following two-stage procedure.

In the first stage, a card is chosen with probability *T* (0 < *T* < 1) to say "Yes" if the person bears *A* and with a probability $(1 - T)$ to go to the second stage of the RRT. If one goes to the second stage, then the instruction is to give an RR according to Warner's scheme. That is, with a probability *p* (0 < *p* < 1) to say "Yes"/"No" about bearing *A* and with probability $(1 - p)$ to say so about *A^c*.

This being truthfully done, $\lambda = T\theta + (1 - T)[p\theta + (1 - p)(1 - \theta)]$ is the probability of a "Yes" response. This obviously yields an unbiased estimator for θ contesting the one under Warner's RRT affording a comparison in the relative efficiencies leading to a condition on the magnitude of *T* one should assign it. This issue has been contested by Moors (1997). If heeding Nayak's (1994) recommendation (1) efficiency in estimation and (2) the level of privacy being protected by an RRT are simultaneously put to a judicious contention then interesting considerations emerge as we shall soon see in brief. Mangat and Singh (1990) add an algebraic elaboration of allowing a possibility of a partial lack of truthfulness in a respondent's report. Such a situation was also visualized by Greenberg et al. (1969) in the context of Warner's (1965) basic scheme as was discussed by Chaudhuri and Mukerjee (1988). Mangat et al. (1992) introduced the revised version of Simmons' RRT described by Greenberg et al. (1969) with a known proportion γ(0 < γ < 1) in the community bearing an innocuous characteristic *B* unrelated to *A* asking a person sampled by SRSWR in any of the *n* draws to say "Yes" if bearing *A* and to apply Simmons's device to say "Yes" or "No" in a truthful way with probability *p* to say "Yes" if bearing *A* or with probability $(1 - p)$ to say "Yes" if bearing *B* or "No" truthfully in the contrary case. Then the probability of a "Yes" answer by Simmons's method or by the above revised method is, respectively

$$\lambda = p\theta + (1 - p)\gamma \quad \text{and} \quad \lambda' = \theta + (1 - \theta)(1 - p)\gamma.$$

Resulting unbiased estimators for θ are then compared in respect to their variances with no considerations of their "privacy protection" levels.

Mangat (1994) gave the under-noted simplification on that by Mangat and Singh (1990) and compared the variances of unbiased estimators of θ based on these two and Warner's (1965) device for θ by quite easily derived algebra because SRSWR in n draws was taken as the common scheme of sampling to gather the RRs by the respective RRTs.

The Mangat's (1994) device is to ask a respondent to say "Yes" if bearing A and if not bearing A^c to truthfully report so if warranted by Warner's RR device.

Then, the probability of a "Yes" response is

$$\lambda = \theta + (1 - \theta)(1 - p) = (1 - p) + p\theta,$$

yielding an unbiased estimator

$$\hat{\theta} = \frac{1}{p}(\hat{\lambda} - 1 + p),$$

writing $\hat{\lambda}$ as the sample proportion of "Yes" responses.

Easy efficiency comparison is then reported covering it vis-à-vis Warner's (1965) and Mangat and Singh contestants in 1990. What happens in case of "less than 100 percent truthful reporting" is also discussed by Mangat (1994).

As touched briefly in Chapter 3, Mangat et al. (1995) considered rival estimation procedures versus Warner's treatment noting the RRs from distinct respondents in an SRSWR in n draws permitting only one RR from each by Warner's device. They consider the following four strategies.

Strategy 1: SRSWR in n draws. A person selected on any draw implements Warner's RR device leading to the unbiased estimator

$$\hat{\theta}_1 = \frac{(n'/n) - 1 + p}{2p - 1}, \quad p \neq \frac{1}{2};$$

here n' is the number of "Yes" responses observed with each person every time selected says "Yes" about bearing A with probability p. Warner (1965) has given its variance as

$$V_1(\hat{\theta}_1) = \frac{\theta(1 - \theta)}{n} + \frac{p(1 - p)}{n(2p - 1)^2}$$

Strategy 2: In n draws by SRSWR d distinct persons happen to be selected; each of these d persons implements Warner's RR device. "Yes" responses emerge from d' of them. These authors estimate θ unbiasedly by

$$\hat{\theta}_d = \frac{(d'/d) - 1 + p}{2p - 1}, \quad p \neq \frac{1}{2}.$$

They work out its variance as

$$V_2(\hat{\theta}_d) = \left[N \, E_1\left(\frac{1}{d}\right) - 1 \right] \frac{\theta(1-\theta)}{N-1} + \frac{p(1-p)}{(2p-1)^2} E_1\left(\frac{1}{d}\right),$$

writing N as the population size and E_1 as the operator for expectation with respect to the probability distribution of d.

Strategy 3: Based on an SRSWOR in n draws formally the same $\hat{\theta}_1$ of Strategy 1 is unbiased for θ with a variance as

$$V_3(\hat{\theta}_3) = \frac{N-n}{N-1} \frac{\theta(1-\theta)}{n} + \frac{p(1-p)}{n(2p-1)^2}.$$

Strategy 4: Based on an SRSWOR in number of draws equal to $E_1(d)$, presumed a positive integer, an unbiased estimator for θ is

$$\hat{\theta}_E = \frac{d'/E_1(d) - 1 + p}{2p - 1}, \quad p \neq \frac{1}{2}$$

with a variance worked out by them as

$$V_4(\hat{\theta}_E) = \left[\frac{N/E_1(d) - 1}{(N-1)} \right] \theta(1-\theta) + \frac{p(1-p)}{E_1(d)(2p-1)^2}.$$

Some algebraic exercises concerning these variances by way of comparing these four strategies have been reported in this work of these five authors. No discussion is included about how far the protection level of a respondent's privacy is.

Mangat et al. (1993) start with the finding of Moors (1971) and produce better results as follows.

Recalling the RRT involving the unrelated model (U-model) of Simmons narrated by Greenberg et al. (1971) for which two independent samples by SRSWR in n_1 and n_2 draws are taken employing the RR device with probability p_1 a respondent is to say "Yes" if bearing A and with probability $(1 - p_1)$ about bearing B and so doing with probabilities p_2 or $(1 - p_2)$ if chosen in the second sample, answering "No" in the contrary case. The unknown proportions bearing A and B respectively are θ_A, θ_B, B denoting an innocuous characteristic unrelated to A. Moors (1971) modifies this by using the second sample exclusively to get DR about B, writing

$$\lambda = \theta_A p_1 + \theta_B + \theta_B(1 - p_1)$$

as the probability of a "Yes" response from the first sample. For simplicity, we shall write p for p_1 here as p_2 is a nonentity with Moors. An unbiased estimator for θ_A is then $\hat{\theta}_A = [(n_1/n) - (1-p)\hat{\theta}_B]/p$. Writing n_1 as the number of "Yes" responses in the first sample and $\hat{\theta}_B$ as the proportion of "Yes" responses about bearing B in the second sample of size $n_2 = n - n_1$ with n as the total size of the two SRSWRs taken together, for an optimal choice of n_1 used in $\hat{\theta}_A$, the latter's variance becomes

$$V(\hat{\theta}_A) = \left[\frac{\lambda(1-\lambda) + (1-p)\sqrt{\theta_B(1-\theta_B)}}{p\sqrt{n}} \right]^2.$$

Mangat et al. (1993) revise the RRT by Moors as follows. A person chosen in the first sample answers "Yes" with a probability T $(0 < T < 1)$ about bearing A and with probability $(1 - T)$ applies the RRT to say with probability p "Yes" about bearing A and with probability $(1 - p)$ about bearing B and says "No" in the contrary cases. As in the RRT by Moors (1971) DR is gathered from the entire second sample about bearing B. Then the probability of gathering a "Yes" response about bearing A is $\alpha = T\theta_A + (1 - T)[p\theta_A + (1 - p)\theta_B]$.

Then, an unbiased estimator for θ_A is

$$\hat{\theta}_A = \frac{\left[\hat{\alpha} - (1 - T)(1 - p)\hat{\theta}_B \right]}{p + T(1 - p)}$$

writing $\hat{\alpha}$ as the proportion of "Yes" RRs in the first sample. Optimizing the choice of n_1, keeping $n_2 = n - n_1$, and writing

$$k = \sqrt{\alpha(1 - \alpha)} + (1 - T)(1 - p)\sqrt{\pi_B p - \pi_B},$$

the optimal variance of $\hat{\theta}_A$ is

$$V_{opt}(\hat{\theta}_A) = \frac{k^2}{\left[n\{p + T(1 - p)\} \right]^2}$$

Calling the $\hat{\theta}_A$ using the optimal values of n_1, the optimum quantity $\hat{\theta}_{A(opt)}$, the authors with detailed algebraic exercises establish the

---

Theorem 9.14

$\hat{\theta}_{A(opt)}$ has a variance smaller than that of the optimal estimator of θ_A given by Moors (1971). The authors Singh et al. (2000) first note the estimator $\hat{\theta}_G = [(1 - p_2)\hat{\theta}_A - (1 - p_1)\hat{\theta}_B/(p_1 - p_2)]$ as given by Greenberg et al. (1969) for

θ_A in U-model with unknown sensitive θ_A and innocuous θ_B with B unrelated to A get the variance

$$V_G(\hat{\theta}_G) = \frac{\left[(1 - p_2)^2 \dfrac{\theta_A(1 - \theta_A)}{n_1}\right] + \left[(1 - p_1)^2 \dfrac{\theta_B(1 - \theta_B)}{n_2}\right]}{(p_1 - p_2)^2}$$

Then they note that if the two independent SRSWR's in n_1, n_2 draws be instead both SRSWOR's then Kim (1978) has derived for θ_A an unbiased estimator $\hat{\theta}_k$ which is same as $\hat{\theta}_G$ but has the variance

$$V_k(\hat{\theta}_k) = V_k(\hat{\theta}_G) = V_G(\hat{\theta}_G) - A(\underline{p}) - B(\underline{p})$$

when

$$A(\underline{p}) = \frac{(1 - p_2)^2 (n_1 - 1)\{p_2^2 \theta_A(1 - \theta_A) + (1 - p_2)^2 \theta_B(1 - \theta_B)\}}{n_2(N - 1)(p_1 - p_2)^2}$$

and

$$B(\underline{p}) = \frac{(1 - p_1)^2 (n_2 - 1)\{p_2^2 \theta_A(1 - \theta_A) + (1 - p_2)^2 \theta_B(1 - \theta_B)\}}{n_2(N - 1)(p_1 - p_2)^2}.$$

As noted earlier in this section, Moors (1971) recommended DRT for the entire second SRSWR. So, one may be tempted to do the same also in case of Kim's (1978) SRSWORs.

But Singh et al. (2000) entirely disapprove Moor's (1971) device. Their very reasonable explanation for disapprobation is simply that the same person announcing "No" about bearing B directly in the second sample will feel the person's privacy is compromised when required to say "Yes" about A in the first sample if chosen on both. They encounter this dilemma with the following recommendation:

Two independent SRSWOR's in n_1 and n_2 draws are chosen. A person in the first sample is requested to say "Yes" with a probability p_1 if the person bears A and with a probability $(1 - p_1)$ to say "Yes" if the person bears B and "No" in the contrary cases. A person chosen in the second sample answers truthfully about bearing B provided the person is not included in the first sample.

A person chosen in the second sample, if also chosen in the first sample is to follow Greenberg et al.'s (1969) RR device, with p_2 and $(1 - p_2)$ as probabilities to say "Yes" about A and B, respectively.

Let n_{21} be the number of units out of n_2 in the second sample that are also common with the first sample and n_{22} be the numbers that are uncommon, $n_{21} + n_{22} = n_2$.

Let $\hat{\alpha}_1 \equiv$ the proportion of "Yes" answers in the first sample,
$\hat{\alpha}_{21} \equiv$ the proportion of "Yes" answers in the second sample that are common with those in the first and $\hat{\theta}_B$ be the proportion of "Yes" responses about B out of the n_{22} persons in the n_2 of the second sample persons uncommon with those in the first sample.

Then, $\hat{\theta}_{A1} = [\hat{\alpha}_1 - (1 - p_1)\hat{\theta}_B]/p_1$ and $\hat{\theta}_{A2} = [\hat{\alpha}_2 - (1 - p_2)\hat{\theta}_B]/p_2$ are two unbiased estimators of θ_A.

Writing E_2 as the operator for expectation with n_{21}, n_{22} held fixed and E_1 that over their variations they work out the variance of $\hat{\theta}_B$ as

$$V(\hat{\theta}_B) = \left[NE_1\left(\frac{1}{n_{22}}\right) - 1 \right]\frac{\theta_B(1 - \theta_B)}{N - 1}$$

$$\text{Cov}(\hat{\alpha}_1, \hat{\alpha}_2) = \frac{N - n_1}{(N - 1)n_1}\left[p_1 p_2\,\theta_A(1 - \theta_A) + (1 - p_1)(1 - p_2)\theta_B(1 - \theta_B) \right]$$

$$V(\hat{\alpha}_1) = \frac{\lambda(1 - \lambda)}{n_1} - \frac{(n_1 - 1)\theta_B(1 - \theta_B)}{(N - 1)n_1}$$

$$V(\hat{\alpha}_2) = \left[\lambda'(1 - \lambda') + \frac{\theta_B(1 - \theta_B)}{N - 1} \right]E_1\left(\frac{1}{n_{21}}\right) - \frac{\theta_B(1 - \theta_B)}{N - 1}$$

writing $\lambda' = p_2\theta_A + (1 - p_2)\theta_B$.

By dint of heavy algebra and utilization of Kim's (1978) works, they worked out the formulae for $V(\hat{\theta}_{A1})$, $V(\hat{\theta}_{A2})$ and introducing an arbitrarily assignable weight w and proposing a pooled estimator for θ_A as $\hat{\theta}_{Ap} = w\hat{\theta}_{A1} + (1 - w)\hat{\theta}_{A2}$ they further worked out optimal choice of w and hence of the optimal estimator for θ_A as $\hat{\theta}_A(opt) = w(opt)\hat{\theta}_{A1} + (1 - w(opt))\hat{\theta}_B$.

As these involve too many unassignable parametric quantities we are not greatly interested in their reproduction.

A further alternative RRT is also dealt with by Singh et al. (2000) allowing a single SRSWOR split up randomly into two mutually exclusive subsamples, one of which employs Greenberg et al.'s RRT and the other a DR technique including only B. Detailed algebraic results are presented. Those interested in the details of RRTs may refer to their work.

Singh's (2002) generalization of Warner's (1965) and Kuk's (1990) RRTs called by him a technique of "stochastic randomized response model" is quite comprehensively discussed. But we find it too specialized a device beyond a usual practitioner's comprehensibility level. So, we desist from elaborating it further. But it is worth while to note that the parameter p in Warner's is assigned a prior probability distribution and so are the parameters θ_1 and θ_2 a joint prior probability distribution pointing out how this may be cleverly utilized to enhance precision in estimation and in the level of protection in the privacy of a truthful respondent, referring to Kuk's (1990) RRT.

Mangat and Singh (1994) for the first time presented a modification to Warner's (1965) RRT permitting a respondent either to (1) give a DR or (2) an RR by Warner's RRT about bearing a sensitive attribute A or its complement A^c by a "Yes" or "No" response without divulging whether the option (1) or (2) is used. The proportion of "Yes" responses thus received from an SRSWR in n draws is taken as a required estimator for θ, the unknown proportion of people bearing A in the community of N persons.

This estimator is valued as a proper fraction whereas Warner's estimator for θ may take a value outside the closed interval [0, 1].

These authors assume that every respondent, no matter in which draw the person is selected, has an unknown but common probability $T(0 < T < 1)$ of giving out the DR if given an option either to give a DR or an RR by Warner's method. Presumably, they assume that in the community there is an unknown proportion $T(0 < T < 1)$ of people ready to give the DR about bearing A or A^c. This is quite acceptable to us. Since an SRSWR is only allowed we see no problems with this.

These authors work out the bias of their proposed estimator for θ which turns out biased. Also they work out its MSE about θ, note that the MSE decreases with T for every $\theta \neq 1/2$ and also find the relative magnitudes of this MSE and the variance of the Warner's estimator presenting tabulated values of the relative efficiency of the two for various n, T, θ, and p.

Gupta et al. (2002) started with the scrambled RRT for quantitative data of Eichhorn and Hayre (1983) and modified it for an improved estimation of a population mean by allowing option for DR.

SRSWR in n draws forms the basis for generating an RR as a variable z obtained by multiplying the stigmatizing variable value y by a scramble variable x with a known distribution with given mean μ_x, variance σ_x^2 which is distributed independently of y. Thus, $z = xy$ is the RR according to Eichhorn et al. (1983) but $r = x^I y$ is the modified RR according to Gupta et al. (2002). Here $I = 1$ if the option to use the scrambler x is exercised and $I = 0$, if not. This I is a random variable and expectation $E(I) = W$ is interpreted as the "Level of Sensitivity" of the item represented by y.

Gupta et al. (2002) had worked out the expectation of r in terms of means of x, y, and W and derived an unbiased estimator of the mean μ_y of y utilizing the property of SRSWR. They also derived the variance of the estimator and showed by how much this variance is lesser than the variance of the Eichhorn et al.'s (1983) unbiased estimator of μ_y. They further obtained an appropriate estimator for W. Also they presented a large sample test for a null hypothesis about W against both sided alternatives. Obviously this optional RR like the one of Mangat and Singh (1994) is totally different from Chaudhuri et al.'s (1985, 1988, and 2005).

Arnab (2004) begins with an inappropriate sentence "Gupta, et al. (2002) proposed an optional randomized response (ORR) technique based on SRSWR sampling using an unrealistic assumption that the probability of providing the true/randomized response for all the individuals in a population is the same." The present reviewer does not find any unrealistic element here as suspected. In a given community there may legitimately exist a section with an unknown proportion $T(0 < T < 1)$ of people willing to give out a true response on request while the remainder may agree to respond implementing a randomization device explained to the person. In the next sentence, Arnab (2004) expresses "In this paper an alternative ORR has been proposed for an arbitrary sampling design." To discuss what he does next it is needful to first recall Eichhorn and Hayre's (1983) RR device produced from the *i*th person in an SRSWR in *n* draws the RR as

$$r_i = \frac{S_i}{\theta} y_i,$$

obtained on multiplying the true unknown sensitive value y_i by a value S_i generated randomly from the distribution of a "scrambling variable S" having the known mean θ and variance γ^2. The unknown mean μ of the y_i's is then unbiasedly estimated by

$$\hat{\mu} = \frac{1}{n} \sum_{i=1}^{n} r_i$$

Its variance is $V(\hat{\mu}) = 1/n[\sigma^2 + C_\gamma^2(\sigma^2 + \mu^2)]$, writing

$$C_\gamma = \frac{\gamma}{\theta}, \quad (\gamma > 0), \quad \sigma^2 = \frac{1}{N} \sum_{1}^{N} (y_i - \mu)^2$$

For simplicity Gupta et al. (2002) take this θ as unity and for their ORR technique modified on Eichhorn and Hayre (1983) the variance of this $\hat{\mu}$ is

$$V'(\hat{\mu}) = \frac{1}{n}\left[\sigma^2 + W\gamma^2\left(\sigma^2 + \mu^2\right)\right]$$

with W as noted earlier is the unknown probability that any person sampled by SRSWR in *n* draws gives out the DR truthfully. This W is called "the degree of confidentiality." Obviously, $V'(\hat{\mu}) < V(\mu)$ because $W < 1$, taking θ as unity here too. Arnab's (2004) own follow-up of an ORR is briefly discussed below.

   Arnab's (2004) main acceptable complaints against the ORR strategy given by Gupta et al. (2002) are two: (1) It is not applicable to sample selection with unequal probabilities and (2) unbiased estimation of the variance of their proposed unbiased estimator for the population mean of the sensitive variable of interest is unavailable because of the presence of the unknown element W. Their third objection about the common element W for the community however, is not tenable because of the arguments we have already provided.

Let us briefly narrate Arnab's (2004) recommendations to get over the twin issues (1) and (2).

Of his own admission Arnab's (2004) optional RR approach mimics that of Chaudhuri and Mukerjee (1985, 1988) and of Chaudhuri and Saha (2005), (Arnab, 2004, p. 116). For each item variable y he supposes every person i must be one of group G to agree to give an RR r_i by the scrambling technique of Eichhorn and Hayre (1983) or to the complementary group $\bar{G}$ to give out the true value y_i "disclosing" to the enquirer which option is actually exercised.

Denoting by r_i the RR from i and by y_i the true value and by δ_i the indicator $1/0$ as $i \in \bar{G}/G$ the ORR is $\tilde{r}_i = \delta_i y_i + (1 - \delta_i) r_i$ for i in a sample s chosen with probability $p(s)$ according to a design p admitting positive inclusion-probabilities $\pi_i, i \in \cup$ and $\pi_{ij}, i,j \in \cup, i \neq j$.

His proposed unbiased estimator for

$$\mu = \frac{1}{N} \sum_{i=1}^{N} y_i \text{ is } t = \frac{1}{N} \sum_{i \in s} b_{si} \tilde{r}_i,$$

with b_{si}'s free of

$$\underline{Y} = (y_1, ..., y_i, ..., y_N), \quad \underline{R} = (r_1, ..., r_i, ..., r_N),$$

$$\sum_{s \ni i} b_{si} p(s) = 1 \, \forall i, \quad \alpha_i = \sum_{s \ni i} b_{si}^2 p(s),$$

$$\alpha_{ij} = \sum_{s \ni i,j} b_{si} b_{sj} p(s), \quad \sigma_i^2 = VR(\tilde{r}_i),$$

variance of t is

$$V(t) = \sum (\alpha_i - 1) y_i^2 + \sum_{i \neq} \sum_j (\alpha_{ij} - 1) y_i y_j + \sum_1^N \alpha_i \sigma_i^2$$

of which an unbiased estimator is

$$\hat{V}(t) = \sum_{i \in s} d_{si} (\tilde{r}_i)^2 + \sum_{i \neq j} \sum_{\in s} d_{sij} \tilde{r}_i \tilde{r}_j + \sum d_{si}^* \tilde{\sigma}_i^2$$

writing $d_{si}, d_{sij}, d_{si}^*$ free of $\underline{Y}, \underline{R}$,

$$\sum_{s \ni i} d_{si} p(s) = \alpha_i - 1, \quad \sum_{s \ni i,j} d_{sij} p(s) = \alpha_{ij} - 1, \quad \sum_{s \ni i} d_{si}^* p(s) = 1$$

and $\tilde{\sigma}_i^2$ satisfying $E_R(\tilde{\sigma}_i^2) = \sigma_i^2$, $i \in \cup = (1, ..., N)$. For example,

$$d_{si} = \frac{(1-\alpha_i)}{\pi_i}, \quad d_{sij} = \frac{(1-\alpha_{ij})}{\pi_{ij}}$$

and

$$d_{si}^* = \frac{1}{\pi_i}$$

and

$$\tilde{\sigma}_i^2 = \frac{\left[C(1-C)r_i^2 - 2Cr_i\left(\sum_j Q_j q_j\right) + \left\{\sum_j Q_j^2 q_j - \left(\sum_j Q_j q_j\right)^2\right\}\right]}{C},$$

with the constants C, q_j, and Q_j as they occur in the RR device given by Eriksson (1973). Arnab (2004) does not explain how to resolve this mix-up of Eriksson's (1973) RRT with that of Eichhorn and Hayre (1983).

Arnab (2004) illustrated a few specific strategies as special cases which we need not quote here. Importantly he showed the above ORR procedure presented by him as more efficient in terms of variances than the corresponding RR procedures permitting no option to give out the genuine variate values by the respondents.

Pal (2008) provides a clue to the use of Eichhorn and Hayre's (1983) scrambling RR device in estimation of a total and of the variance of the "total-estimator" when a sample s is selected by any unequal probability sampling design. The author further provides a thoroughly novel ORR version of this particular strategy as a novel seminal work.

Let me briefly recapitulate this pioneering work here. This involves considerably complex algebraic manipulations. Our approach to essentially solve the same problem is however, quite different from hers and is appreciably simpler in formulation.

Writing y_i for the sensitive (y) variate value, x_i for the scrambling variable x with known mean θ_1 and variance σ_1^2, u_i that of another variable independent of x with known mean α_1 and variance v_1^2, Pal (2008) supposes, from a sampled person i, the RR is

$$I_i = \frac{y_i x_i}{\theta} + u_i$$

having $E_R(I_i) = y_i + \alpha_1$ and hence $r_i = (I_i - \alpha_i)$ has $E_R(r_i) = y_i$ and

$$V_R(r_i) = y_i^2\left(1 + \frac{\sigma_1^2}{\theta_1^2}\right) + v_1^2 - y_i^2 = E_R(r_i^2) - E_R^2(r_i)$$

and hence

$$\hat{y}_i^2 = \frac{r_i^2 - \gamma_1^2}{1 + (\sigma_1^2/\theta_1^2)}$$

giving $\hat{y}_i^2 = (\sigma_1^2/\theta_1^2) + \gamma_1^2 = \hat{V}_R(r_i)$ an unbiased estimator $\hat{V}_i$ for $V_R(r_i) = V_i$.

The ORR version of it given by Pal (2008) assumes $C_i (0 < C_i < 1 \forall i \in \cup)$ as an unknown probability that the ith person on request may disclose the true sensitive value y_i and with a probability $(1 - C_i)$ may give out the RR by Eichhorn et al.'s (1983) scrambling procedure I_i as described above. Then, the RR is

$$Z_i = y_i \text{ with probability } C_i$$
$$= I_i \text{ with probability } (1 - C_i),$$

$$E_R (Z_i) = y_i C_i + (1 - C_i) E_R (I_i)$$
$$= y_i + (1 - C_i)\alpha_1.$$

Next, Pal (2008) requires an additional RR as Z'_i independently of Z_i in the following way. In addition to x_i, a second scrambling variable x' distributed independently of x with a mean θ_2 and a variance σ_2^2, both known is introduced and also another random variable u' distributed independently of u, x, x' but with known mean α_2 and known variance γ_2^2 is introduced. Writing $I'_i = y_i(x'_i/\theta_2) + u_2$ and introducing a constant p $(0 < p < 1)$, Pal (2008) designs from a sampled i an ORR as

$$Z'_i = y_i \text{ with probability } C_i$$
$$= pI_i + (1 - p) I'_i \text{ with a probability } (1 - C_i), \quad i \in \cup$$

Then

$$E_R (Z'_i) = y_i C_i + (1 - C_i)\left[pE_R(I_i) + (1 - p)E_R(I'_i) \right]$$
$$= y_i + (1 - C_i) \left[p(\alpha_1 - \alpha_2) + \alpha_2 \right].$$

Writing $\alpha'_2 = p(\alpha_1 - \alpha_2) + \alpha_2$, $\alpha'_2 \neq \alpha_1$, it follows that

$$r'_i = \frac{\alpha'_2 Z_i - \alpha_1 Z'_i}{\alpha'_2 - \alpha_1} \quad \text{has } E_R (r'_i) = y_i, \quad i \in \cup,$$
$$V_R (r'_i) = E_R (r'^2_i) - y_i^2 = V_i, \text{ say.}$$

Next, writing

$$\hat{C}_i = 1 - \left(\frac{Z_i - Z_i'}{\alpha_i - \alpha_2'} \right),$$

one has $E_R(\hat{C}_i) = C_i$. Further,

$$E_R(Z_i^2) = y_i^2 + (1 - C_i)y_i^2 \frac{\sigma_1^2}{\theta_1^2} + (1 - C_i)(\alpha_1^2 + \gamma_1^2) + 2(1 - C_i)y_i\alpha_1$$

and

$$E_R\left(Z_i'^2\right) = y_i^2 + \left(1 - C_i\right)y_i^2 \, \phi$$
$$+ (1 - C_i)\left[p\left(\alpha_1^2 + \gamma_1^2\right) + (1 - p)\left(\alpha_2^2 + \gamma_2^2\right) \right]$$
$$+ 2(1 - C_i)y_i\alpha_2'.$$

Writing

$$\phi = \frac{p\sigma_1^2}{\theta_1^2} + \frac{(1 - p)\sigma_2^2}{\theta_2^2},$$

$$E_R\left(Z_iZ_i'\right) = y_i^2 + \frac{\sigma_1^2}{\theta_1^2}y_i^2\left(1 - C_i\right)p + (1 - C_i)\left[p\left(\alpha_1^2 + \gamma_1^2\right) + (1 - p)\alpha_1\alpha_2 \right]$$
$$+ \left[\alpha_1\left(1 + p\right) + \alpha_2\left(1 - p\right) \right]y_i\left(1 - C_i\right).$$

Writing $\psi = \sigma_1^2/\theta_1^2$, $\alpha_3' = \alpha_1(1 + p) + \alpha_2(1 - p)$ and stipulating that $\phi \neq \psi$, it follows that

$$\hat{y}_i'^2 = \frac{\left(\alpha_3'\phi - \alpha_2'p\psi\right)\left[\alpha_2'\left(Z_i^2 - \hat{M}\right) - \alpha_1\left(Z_i'^1 - \hat{M}'\right)\right]}{\left(\alpha_2' - \alpha_1\right)\left(\alpha_3'\phi - \alpha_2'\phi\psi\right) - \left(\alpha_3' - \alpha_2'\right)\left(\psi\alpha_2' - \phi\alpha_1\right)}.$$

Writing

$$\hat{M} = (1 - \hat{C}_i)(\alpha_1^2 + \gamma_1^2),$$
$$\hat{M}' = (1 - \hat{C}_i)\left[p(\alpha_1^2 + \gamma_1^2) + (1 - p)(\alpha_2^2 + \gamma_2^2) \right] \text{ and}$$
$$\hat{M}'' = (1 - \hat{C}_i)\left[p(\alpha_1^2 + \gamma_1^2) + (1 - p)\alpha_1\alpha_2 \right], \quad V_R(r_i),$$

has an unbiased estimator as $\hat{V}_i' = r_i'^2 - \hat{y}_i'^2$.

Estimation of $Y = \Sigma_{i=1}^{N} y_i$ by $e = \Sigma_{i \in s} b_{si} r_i'$ and of $V(e)$ by either

$$v_1(e) = \hat{V}_p\left(\sum_{i \in s} b_{si} y_i\right)\bigg|_{\underline{Y} = (r_1', \dots, r_i', \dots r_N')} + \sum_{i \in s} \hat{V}_i' b_{si} \quad \text{or by} \quad v_2(e)$$

$$= V_p\left(\sum_{i \in s} y_i b_{si}\right)\bigg|_{\underline{Y} = (r_1', \dots, r_i', \dots r_N')} + \sum_{i \in s} \hat{V}_i'(b_{si}^2 - d_{si})$$

follows easily with b_{si}, d_{si} as treated in the case of Gupta et al.'s (2002) works above.

Pal (2008) illustrated applications with specific strategies demonstrating numerically that the ORR procedures achieve reduced variances compared to the corresponding compulsory randomize response (CRR) procedures.

Singh and Joarder (1997) have the following solution to the problem posed by Gupta et al. (2002) quite earlier than the latter. They consider Eichhorn and Hayre's (1983) RR as $z_i = y_i s_i/\theta$ from a sampled person as i with true value y_i in an SRSWR in n draws with s_i as the values for a scrambling variable s with mean θ and variance γ^2. Writing μ and σ^2 for the population mean and variance for the stigmatizing variable y their proposed estimator for μ is

$$e = \left(\frac{1}{n\theta}\right)\sum_{i=1}^{n} z_i \quad \text{and} \quad V(e) = \frac{1}{n}\left[\sigma^2\left(\frac{\gamma}{\theta}\right)^2(\sigma^2 + \mu^2)\right]$$

Their ORR from the ith person is $z_i' = y_i$ with probability W, and $y_i x_i/\theta$ with probability $(1 - W)$ as in Gupta et al.'s (2002).

Their unbiased estimator for μ is then $e' = (1/n)\Sigma_{i=1}^{n} z_i'$ for which obviously $E(e') = \mu(y_i x_i/\theta)$.

Also, they show with a simple algebra that

$$V(e') = \frac{1}{n}\left[\sigma^2\left(\frac{\gamma}{\theta}\right)^2(\sigma^2 + \mu^2)(1 - W)\right]$$

Most importantly they show that $s_y^2 = 1/(n - 1)\Sigma_{i=1}^{n}(z_i' - e')^2$ has $E(s_y^2) = V(e')$, that is, s_y^2 is an exactly unbiased estimator for $V(e)$. Further, they show that

$$\frac{V(e)}{V(e')} \times 100 = \frac{\left(\dfrac{\sigma}{\mu}\right)^2 + \left(\dfrac{\gamma}{\theta}\right)^2\left(1 + \dfrac{\sigma^2}{\mu^2}\right)}{\left(\dfrac{\sigma}{\mu}\right)^2 + \left(\dfrac{\gamma}{\theta}\right)^2\left(1 + \dfrac{\sigma^2}{\mu^2}\right)(1 - W)},$$

and since $0 \le W \le 1$, it follows that the ORR version e' of the Eichhorn and Hayre's (1983) RRT is more efficient than e, the original RR estimator of Eichhorn and Hayre (1983). This work clearly takes away a major portion of the credit of Gupta et al.'s (2002) ORR approach.

A few more RR strategies in the literature after 1988:

Singh and Joarder (1999) have an "unknown repeated trial model" to improve upon Warner's (1965) model. Here Warner (1965) is the starting point in an SRSWR in n draws for a sampled person who is to repeat the trial if in the first trial the RR device does not produce a "Match" for his own trait A/A^c and without disclosing this to the enquirer repeats the Warner RR device to give his truthful A/A^c status in accordance with the outcome of the RR device after the second repeat. With p as the probability that an A-marked card is drawn rather than one marked A^c for a truthful "Yes" response, according to this scheme is $\lambda = \theta[p + (1 - p)p] + (1 - \theta)(1 - p) = Prob$ ("Yes").

Letting $\hat{\lambda}_1$ as the sample proportion of "Yes" responses as recorded

$$\hat{\theta} = \frac{\hat{\lambda}_1 - (1 - p)}{2p - 1 + p(1 - p)}$$

is an unbiased estimator for θ. The variance of $\hat{\theta}$ is given by them as

$$V\left(\hat{\theta}\right) = \frac{\theta(1 - \theta)}{n} + \frac{p(1 - p)}{n\left[(2p - 1) + p(1 - p)\right]^2} - \frac{\theta p(1 - p)}{n\left[(2p - 1) + p(1 - p)\right]}.$$

An unbiased estimator for $V(\hat{\theta})$ is

$$\hat{V}(\hat{\theta}) = v(\hat{\theta}) = \frac{\hat{\lambda}_1(1 - \hat{\lambda}_1)}{(n - 1)\left[(2p - 1) + p(1 - p)\right]^2}.$$

Singh and Joarder (1997) have shown that their estimator has a smaller variance than the one given by Warner (1965) for every $p > 1/2$ and have presented a table showing a numerical comparison between the two with an appreciable advantage in their favor.

For repeated RRs by design, Franklin's (1989a,b) and Singh and Singh's works (1992, 1993) from SRSWR data and Pal's (2002) generalization to unequal probability samples.

To estimate $\theta = (1/N)\sum_1^N y_i$, with y_i as a sensitive attribute's 1/0 values of a stigmatizing characteristic A that an ith person ($i \in \cup = (1, \ldots, i, \ldots N)$) bears, Franklin (1989a,b) recommends gathering from each person i chosen in an SRSWR in n draws observations $x_{i1}, \ldots, x_{ij}, \ldots, x_{ik}$ on K independently distributed variables x_j with means μ_{1j} and variances σ_{1j}^2 if i bears A_j. On the other hand if i bears A^c, the values reported by i are $u_{i1}, ..., u_{ij}, ..., u_{ik}$ on the variables u_j distributed independently of x_j $(j = 1, ..., K)$ and of each other with means μ_{2j} and variances σ_{2j}^2 $(j = 1, ..., K)$. These $\mu_{1j}, \mu_{2j}, \sigma_{1j}(>0), \sigma_{2j}(>0)$ are all supposedly known.

Writing $y_i = 1/0$ if i bears A/A^c, the RR data at hand will be z_{ij} as either x_{ij} or y_{ij} for $(j = 1, ..., K)$ not knowing which one of course. Thus,

$$z_{ij} = y_i x_{ij} + (1 - y_i)u_{ij}, \quad j = 1, ..., K$$

for i in a sample s drawn. Since s is an SRSWR in n draws, one gets

$$E_R(z_{ij}) = \theta\mu_{1j} + (1 - \theta)\mu_{2j}$$

and variance

$$V_R(z_{ij}) = \theta\sigma_{1j}^2 + (1 - \theta)\sigma_{2j}^2 + \theta(1 - \theta)(\mu_{1j} - \mu_{2j})^2$$

and covariances

$$C_R(z_{ij}, z_{ij}') = \theta(1 - \theta)(\mu_{1j} - \mu_{2j})(\mu_{1j'} - \mu_{2j'})$$

$j = 1, ..., K; j'(\neq j) = 1, ..., K; \quad i \in \cup = (1, ..., N)$.

Franklin (1989a,b) has given us two estimators t_1 and t_2, as described below:

$$t_1 = \frac{1}{n}\sum_{i=1}^{n}\hat{\theta}_i; \quad \hat{\theta}_i = \frac{Z_{i0} - m_2}{m_1 - m_2}; \quad Z_{i0} = \sum_{j=1}^{K}Z_{ij},$$

$$m_r = \sum_{j=1}^{K}\mu_{rj}; \quad r = 1, 2; \quad m_1 \neq m_2,$$

assumed; then, $E_R(t_1) = \theta$ and

$$V_R(t_1) = \frac{\theta(1 - \theta)}{n} + \frac{\theta(\sigma_1^2 - \sigma_2^2) + \sigma_2^2}{n(m_1 - m_2)^2},$$

writing $\sigma_{1r}^2 = \Sigma_{j=1}^{K}\sigma_{rj}^2$, $r = 1, 2$ the other estimator for θ proposed by Franklin (1989a,b) is

$$t_2 = \sum_{j=1}^{K}W_j\left[\frac{(\bar{Z}_{oj} - \mu_{2j})}{(\mu_{1j} - \mu_{2j})}\right], \quad \bar{Z}_{oj} = \frac{Z_{oj}}{n}, \quad Z_{oj} = \sum_{i=1}^{n}Z_{ij},$$

with W_j as arbitrarily assignable weights $(0 < W_j < 1, \Sigma_1^K W_j = 1)$. It is checked that $E(t_2) = \theta$ and

$$V(t_2) = \frac{\theta(1-\theta)}{n} + \frac{1}{n}\sum_{j=1}^{K} \frac{W_j^2\left[\theta\left(\sigma_{1j}^2 - \sigma_{2j}^2\right) + \sigma_{2j}^2\right]}{\left(\mu_{1j} - \mu_{2j}\right)^2}.$$

For the choice

$$W_j = \frac{\left|\mu_{1j} - \mu_{2j}\right|}{\sum\limits_{j=1}^{K}\left|\mu_{1j} - \mu_{2j}\right|} = \frac{D_j}{D}, \text{ say,}$$

t_2 achieves its optimal form

$$t_{2opt} = \frac{1}{D}\sum_{1}^{K} \frac{D_j}{\left(\mu_{1j} - \mu_{2j}\right)}\left(\bar{Z}_{oj} - \mu_{2j}\right)$$

with the variance $V_R(t_2)$ minimized at

$$V\left(t_{2opt}\right) = \frac{\theta(1-\theta)}{n} + \frac{1}{nD^2}\left[\theta\left(\sigma_1^2 - \sigma_2^2\right) + \sigma_2^2\right].$$

These works by Franklin (1989a,b) have been modified by Singh and Singh (1992) with the extension that a person labeled i on request, with a probability P_r (assigned by the enquirer such that $0 < P_r < 1$, $r = 1, 2, 3$, such that $P_1 + P_2 + P_3 = 1$) realizes K numbers x_{rij} $j = 1, ..., K$) if i bears A or the K numbers y_{rij} $j = 1, ..., K$) if i bears A^c from distribution independent of each other, respectively with $(\mu_{1j}, \sigma_{rij}^2)$ and $(\mu_{2j}, \sigma_{rij}^2)$ as the means and variances, all known.

Actually, observable RRs are then denoted by L_{ij} for $j = 1, ..., K$ and $i \in \cup$. Then

$$E_R(L_{ij}) = \theta(\mu_{1j} - \mu_{2j}) + \mu_{2j}$$

and

$$V_R(L_{ij}) = \theta\left[\sum_{r=1}^{3} P_r(\sigma_{1j}^2 - \sigma_{2j}^2) + \sum_{r=1}^{3} P_r\sigma_{rij}^2\right] + \theta(1-\theta)(\mu_{1j} - \mu_{2j})^2$$

and

$$C_r(L_{ij}, L_{ij'}) = \theta(1-\theta)(\mu_{1j} - \mu_{2j})(\mu_{1j'} - \mu_{2j'})$$
$$j \in \cup_j \ j = 1, \dots, K; \ j, j'(j \neq j') = 1, \dots, K.$$

Writing $\bar{L}_{i0} = (1/K) \, \Sigma_{j=1}^K L_{ij}$ they take

$$e_1 = \frac{1}{n(m_1 - m_2)} \left[K \sum_1^n \bar{L}_{i0} - m_2 \right]$$

to estimate θ because $E_R(e) = \theta$. They work out

$$V(e_1) = \frac{\theta(1-\theta)}{n} + \frac{\left[\theta \sum_{r=1}^3 P_r \left(\sum_{j=1}^K (\sigma_{r1j}^2 - \sigma_{r2j}^2) \right) + \sum_{r=1}^3 P_r \sum_{j=1}^K \sigma_{r2j}^2 \right]}{n(m_1 - m_2)^2}.$$

Further, they propose a second estimator

$$e_{2w} = \sum_{j=1}^K W_j \left[\frac{1}{(\mu_{1j} - \mu_{2j})} \left\{ \frac{1}{n} \sum_{i=1}^n L_{ij} - \mu_{2j} \right\} \right],$$

since $E_R(e_{2w}) = \theta$ for arbitrary choices of $W_j(0 < W_j < 1, \Sigma_1^K W_j = 1)$, working out

$$V(e_{2w}) = \frac{\theta(1-\theta)}{n} + \frac{1}{n} \sum_1^K W_j^2 \frac{\left[\theta \sum_r P_r (\sigma_{1j}^2 - \sigma_{2j}^2) + \sum_r P_r \sigma_{r2j}^2 \right]}{(\mu_{1j} - \mu_{2j})^2}.$$

For the choice

$$W_j = \frac{|\mu_{1j} - \mu_{2j}|}{\sum_j |\mu_{1j} - \mu_{2j}|} = \frac{D_j}{D},$$

say, e_{2w} is optimized at

$$e_{2opt} = \frac{1}{D} \sum_1^K D_j \left[\frac{1}{(\mu_{1j} - \mu_{2j})} \left(\frac{1}{n} \sum_{i=1}^n L_{ij} - \mu_{2j} \right) \right]$$

with the minimized variance as

$$V\left(e_{2opt}\right) = \frac{\theta(1-\theta)}{n} + \frac{1}{nD^2}\left[\theta\sum_r P_r \sum_j \left(\sigma_{r1j}^2 - \sigma_{r2j}^2\right)\sum_r P_r \sum_j \sigma_{r2j}^2\right].$$

Singh and Singh (1993), as a matter of fact modified the work of Singh and Singh (1992) only replacing $r = 3$ by $r = 2$ above creating the simplification by taking $P_1 = T$, $P_2 = 1 - T$ so that $P_1 + P_2 = 1$ and deriving the optimal e_{1opt} and e_{2opt} changed into e_{1T} and e_{2T} respectively, so as to derive

$$V\left(e_{1T}\right) = \frac{\theta(1-\theta)}{n} + \frac{\theta\sum_j \sigma_{21j}^2 + (1-\theta)\sum \sigma_{22j}^2}{n(m_1 - m_2)^2}$$

$$+ \frac{T}{n(m_1 - m_2)^2}\left[\theta\left(\sum \sigma_{1j}^2 - \sum \sigma_{2j}^2\right) + (1-\theta)\right.$$

$$\left. \times \sum_j \left(\sigma_{12j}^2 - \sigma_{22j}^2\right)\right]$$

$$V\left(e_{2T}\right) = \frac{\theta(1-\theta)}{n} + \frac{1}{nD^2}\left[\theta\sum \sigma_{21j}^2 + (1-\theta)\sum \sigma_{22j}^2\right]$$

$$+ \frac{T}{nD^2}\left[\theta\sum\left(\sigma_{11j}^2 - \sigma_{21j}^2\right) + (1-\theta)\sum\left(\sigma_{12j}^2 - \sigma_{12j}^2\right)\right].$$

Pal (2002), in the unpublished thesis, extended these works of Franklin (1989a,b) and Singh and Singh (1992, 1993) to derive corresponding results for general unequal probability sampling schemes even without replacements breaking the shackles of SRSWR in n draws alone without disturbing the replicated RRs derived from each sampled person no matter how selected. Pal's thesis replicating the approach of Chaudhuri (2001) need not be reproduced here.

Incidentally, neither Franklin (1989a,b) nor Singh and Singh (1992, 1993) mentioned anything about estimation of the variances of the respective estimators for θ they provided. But following Chaudhuri's (2001) approach it was a simple matter for Pal (2002) to present usual variance estimators for the author's revised estimators for θ. We also omit the author's formulae as they do not offer any novelty in approach.

Bhargava and Singh (1999, 2003) compared efficiencies of RRT's under common privacy protection. They gave the variances of unbiased estimators for θ, the proportion bearing the sensitive attribute A in a community of N people either with A or its compliment A^c as given by RRTs by

Warner (1965), Mangat and Singh (1990) and Mangat (1994) under two circumstances 1 and 2.

All the three RRTs are applied to individuals chosen by SRSWR in n draws. In Warner's (1965) RRT, let p_1, $(1 - p_1)$ denote proportion of cards marked A, A^c, $(0 < p_1 < 1, p_1 \neq 1/2)$ and $\hat{\lambda}_1$ the proportion in the sample saying "Yes" about bearing A, giving

$$\hat{\theta}_W = \frac{\hat{\lambda}_1 - (1 - p_1)}{2p_1 - 1}$$

as the unbiased estimator of θ with variance

$$V\left(\hat{\theta}_w\right) = \frac{\theta(1 - \theta)}{n} + \frac{p_1(1 - p_1)}{(2p_1 - 1)^2}. \tag{9.6}$$

In Mangat et al.'s (1990) RRT, with probability $T(0 < T < 1)$ a respondent is to give out the truth about bearing A and with probability $(1 - T)$ give an RR following Warner's (1965) device, with p_1 changed to p_2 such that $T \neq (1 - 2p_2)/[2(1 - p_2)]$. Their unbiased estimator for θ is

$$\hat{\theta}_{Ms} = \frac{\hat{\lambda}_2 - (1 - T)(1 - p_2)}{(2p_2 - 1)2T(1 - p_2)} \quad \text{with variance as}$$

$$V(\hat{\theta}_{Ms}) = \left\{ \frac{\theta(1 - \theta)}{n} \right\} + \frac{(1 - T)(1 - p_2)\left[T(1 - T)p_2 \right]}{n\left[2\{(T + p_2(1 - T)\} - 1 \right]^2}. \tag{9.7}$$

In Mangat's (1994) RRT a respondent is to say "Yes" if the person bears A but if not, is to give the RR by Warner's (1965) RR device, with p_1 now replaced by p_3, $(0 < p_3 < 1)$. Writing $\hat{\lambda}_3$ as the sample proportion of "Yes" responses his estimator for θ is $\hat{\theta}_M = [\hat{\lambda}_3 - (1 - p_3)]/p_3$ with the variance as

$$V(\hat{\theta}_M) = \left\{ \frac{\theta(1 - \theta)}{n} \right\} + \frac{(1 - \pi)(1 - p_3)}{np_3}.$$

In Equation 9.6, Bhargava et al. (1999) employ the concept of "privacy protection" following Lanke's (1976) prescription. Observing the "design probabilities" $P("Yes"/A)$, $P("Yes"/A^c)$, $P("No"/A)$, and $P("No"/A^c)$ the "revealing probabilities" by Bayes' theorem are

$$P(A/\text{"Yes"}) = \frac{\theta P(\text{"Yes"}/A)}{\theta P(\text{"Yes"}/A) + (1-\theta)P(\text{"Yes"}/A^c)} \quad \text{and}$$

$$P(A/\text{"No"}) = \frac{\theta P(\text{"No"}/A)}{\theta P(\text{"No"}/A) + (1-\theta)P(\text{"No"}/A^c)}$$

Lanke's (1976) measure of "privacy protection" is

$$P = \max\,[P(A/\text{"Yes"}), P(A/\text{"No"})] \tag{9.8}$$

Calculating the values of this P-parameter for the above three RRTs, Bhargava et al. (1999) first fix these values at par and work out relations among the parameters involved in the corresponding three variance formulae and then compare among these three variances in terms of the parameters involved. Thus, first they work out

$$P_1 = T\,(1-T)\,p_2 \tag{9.9}$$

to keep the P-values for Warner's and Mangat and Singh's RRTs at par. Equation 9.9 leads the $V(\hat{\theta}_W)$ to reduce to

$$V'(\hat{\theta}_W) = \left\{\frac{\theta(1-\theta)}{n}\right\} + \frac{\left[T + (1-T)p_2(1-T)(1-p_2)\right]}{n\left[2\{T+(1-T)p_2\}-1\right]^2}. \tag{9.10}$$

As Equation 9.10 equals Equation 9.9, they establish Theorem 9.16. Warner's (1965) and Mangat and Singh's (1990) RRT's are equally efficient when constrained to protect a respondent's privacy on an equal footing.

In order to keep the value of the P-parameter at par, for Warner's (1965) and Mangat's (1994) RRTs they work out the required condition as

$$p_1 = \frac{1}{(2-p_3)}.$$

Bhargava et al. (1999) show that the imposition of this condition leads to the $V(\hat{\theta}_W)$ reducing to

$$V''(\hat{\theta}_W) = \left\{\frac{\theta(1-\theta)}{n}\right\} + \frac{1-p_3}{np_3^2}. \tag{9.11}$$

With some simple algebra they show that $V(\hat{\theta}_M) < V''(\hat{\theta}_W)$ for every permissible p_3 and θ. Hence their

Theorem 9.15

Mangat's (1994) RRT achieves a higher efficiency level than the Warner's (1965) if they are both constrained to be at par in protecting a respondent's privacy.

In Equation 9.6, Bhargava et al. follow the recommendation given by Leysieffer and Warner (1976) in protecting a respondent's privacy. Leysieffer and Warner (1976) need

$$g(\text{"Yes"}|A) = \frac{P(\text{"Yes"}|A)}{P(\text{"Yes"}|A^c)}, \quad g(\text{"Yes"}|A^c) = \frac{1}{g(\text{"Yes"}|A)}$$

called the "jeopardy functions" as also

$$g(\text{"No"}|A) = \frac{P(\text{"No"}|A)}{P(\text{"No"}|A^c)}, \quad g(\text{"No"}|A^c) = \frac{1}{g(\text{"No"}|A)}$$

to demand that $g(\text{"Yes"}|A)$ should be greater than unity and also $g(\text{"No"}|A^c)$ be greater than unity at the same time. In fact, for the sake of "privacy protection" they recommend taking $g(\text{"Yes"}|A)$ as large as possible, say $K_1 > 1$ and simultaneously $g(\text{"No"}|A^c)$ also as large as possible, say $K_2 > 1$. So, the design parameters $P(\text{"Yes"}|A)$, $P(\text{"No"}|A)$, $P(\text{"Yes"}|A^c)$, and $P(\text{"No"}|A^c)$ should be rationally worked out in terms of these parameters K_1 and K_2.

For Warner's (1965) RRT, they check that

$$g_W(\text{"Yes"}|A) = \frac{p_1}{1 - p_1}, \quad g_W(\text{"No"}|A^c) = \frac{p_1}{1 - p_1}$$

Taking $K_1 = K_2 = K > 1$, they check that the maximization of $g_w(\text{"Yes"}|A)$ and $g_w(\text{"No"}|A^c)$ leads to the choice of the design

$$\frac{p_1}{1 - p_1} = K \quad \text{that is,} \quad p_1 = \frac{K}{K + 1}.$$

This gives

$$V(\hat{\theta}_W) = \frac{\theta(1 - \theta)}{n} + \frac{K(K - 1)^2}{n} \tag{9.12}$$

For Mangat and Singh's (1990) RRT they note

$$g_{Ms}(\text{"Yes"}|A) = \frac{T + (1 - T)p_2}{(1 - T)(1 - p_2)} = g_{Ms}(N|A^c)$$

and work out

$$p_2 = \frac{K(1-T)-T}{(1-T)(1+K)},$$

leading to the formula, for

$$V(\hat{\theta}_W) = \frac{\theta(1-\theta)}{n} + \frac{K(K-1)^2}{n}. \tag{9.13}$$

Since Equation 9.12 equals Equation 9.13, Bhargava et al. formulated Theorem 9.16. Warner's (1965) RRT is as efficient as the one of Mangat and Singh (1990) if they are constrained to achieve a common level of privacy protection in terms of the jeopardy functions prescribed by Leysieffer et al. (1976) and Bhargava et al. (2003) further work out $P(\text{"Yes"}|A) = 1, P(\text{"No"}|A) = 0,$ $P(\text{"Yes"}|A^c) = 1 - p_3$ and $P(\text{"No"}|A^c) = p_3$ and hence $g_m(\text{"Yes"}|A) = 1/(1 - p_3)$ and $g_m(\text{"No"}|A^c) = \alpha$ for Mangat's (1994) RRT. This leads to the prescription

$$p_3 = \frac{K-1}{K}. \tag{9.14}$$

Utilizing Equation 9.14, the variance formula for $\hat{\theta}_M$ is

$$V(\hat{\theta}_M) = \frac{\theta(1-\theta)}{n} + \frac{(1-\theta)(K-1)^2}{n}. \tag{9.15}$$

Comparing Equation 9.15 with Equation 9.12 follows Bhargava et al.'s (2003):

Theorem 9.16

Mangat's (1994) RRT yields a smaller variance than Warner's (1965) RRT if they are constrained to achieve the same level of protected privacy as per Leysieffer and Warner's (1976) recommendation.

Sarjinder Singh, Stephen Horn, Ravindra Singh, and N.S. Mangat

Singh et al. (2003) modified Horvitz et al.'s (1967) unrelated response model (UY model) of Simmons with known proportion θ_B of people bearing the innocuous characteristic B unrelated to A as described by Greenberg et al. (1969) by allowing cards of three types with proportion p_1 of A, p_2 of B and p_3 that are blank $(p_1 + p_2 + p_3 = 1)$. A respondent in an SRSWR in n draws is to say "Yes" if the respondent's features match on drawing a card of type A or B but otherwise is to say "No." Assuming θ_B as known

$$\hat{\theta}_A = \frac{\left(\dfrac{n'}{n}\right) - (1 - p_1)\theta_B}{p_1}$$

is an unbiased estimator of θ, the proportion bearing A in the community of N people; here n' is the number of "Yes" replies. Its variance is

$$V(\hat{\theta}_A) = \frac{\theta(1-\theta)}{n} + \frac{\theta(1 - p_1 - 2p_2\theta_B)}{np_1} + \frac{p_2\theta_B(1 - p_2\theta_B)}{np_1^2}$$

An unbiased estimator of

$$V(\hat{\theta}_A) \text{ is } v(\hat{\theta}_A) = \frac{\dfrac{n'}{n}\left(1 - \dfrac{n'}{n}\right)}{(n-1)p_1^2}.$$

They have certain findings along with comments on the relative performances of their revised estimator versus the one given by Greenberg et al. (1969).

Sarjinder Singh, Stephen Horn, and Sadeq Chowdhury

These authors address the twin problem of estimating (1) the proportion θ of people with a stigmatizing characteristic, A and simultaneously (2) the unknown mean of a variable directly related to that stigmatizing characteristic. For this they need two SRSWRs in n_1 and n_2 draws employing two RR devices in independent manner.

A respondent i in the first SRSWR gives an RR as

$$Z_{1i} = X_i \text{ with probability } \theta$$

$$= R_{1i} \text{ with probability } 1 - \theta.$$

A respondent j in the second SRSWR gives, on request, the RR as

$$Z_{2j} = X_j \text{ with probability } \theta$$

$$= R_{2j} \text{ with probability } 1 - \theta.$$

Here θ is the unknown proportion in a population bearing A; X is a stigmatizing variable with an unknown mean μ_x. Further R_1 is a random variable with a known mean, variance (θ_1, σ_1^2) and so is R_2 with known mean, variance

(θ_2, σ_2^2), R_1, R_2 are independent of each other. Writing $\bar{Z}_1, \bar{Z}_2$ as the two sample means of Z_{1i}'s and Z_{2j}'s respectively,

$$\hat{\theta} = 1 - \frac{\bar{Z}_1 - \bar{Z}_2}{\theta_1 - \theta_2}, \quad \text{with } \theta_1 \neq \theta_2$$

is proposed to unbiasedly estimate θ, and

$$\hat{\mu}_x = \frac{\bar{Z}_2\theta_1 - \bar{Z}_1\theta_2}{\left(\bar{Z}_2 - \theta_2\right) - \left(\bar{Z}_1 - \theta_1\right)}$$

is their proposed estimator for μ_x. They have given formulae for var($\hat{\theta}$) and for $V(\hat{\mu}_x)$ and worked out formulae for appropriate sample sizes n_1 and n_2 in terms of the related parameters involved. But they have not given any variance estimators.

The authors gave an alternative RR approach for the same twin problems as above to put it in a position to render comparability with Warner's device. We omit the details here.

Chang and Liang

In this work, Chang and Liang (1996) apply Mangat and Singh's (1990) modification on Simmons's unrelated innocuous response model described by Horvitz et al. (1967) and Greenberg et al. (1969) rather than on Warner's (1965) original RR device itself.

As most of the contributors to RRT, these two also restrict to SRSWR alone. Allowing only one feature B, which is unrelated to the sensitive attribute A with unknown proportion θ in a community but is innocuous itself with a known proportion θ_B say in the same community where only one SRSWR in n draws is envisaged. This yields

$$\hat{\theta} = \frac{\left[\left(\dfrac{n'}{n}\right) - (1 - p)\theta_B\right]}{p}$$

as the unbiased estimator for θ based on n' "Yes" responses using a pack of cards with a given proportion p ($0 < p < 1$) asking for a response about A and the rest demanding "Yes"/"No" truthful response about B:

$$V(\hat{\theta}) = \frac{\theta(1-\theta)}{n} + \frac{(1-p)^2 \dfrac{n'}{n}\left(1 - \dfrac{n'}{n}\right) + p(1-p)(\theta + \theta_B - 2\theta\theta_B)}{np^2}$$

The probability of a "Yes" response here is

$$\phi = p\theta + (1 - p)\theta_B,$$

assuming truthful response only. If one bearing A tells the truth only with a probability T', then it is

$$\phi' = p\theta T' + (1 - p)\theta_B.$$

Then $\hat{\theta}$ is biased with a bias $B = (T' - 1)\theta$.

Mangat and Singh's (1990) modification is described earlier in this chapter. In Chang and Liang's (1998) present work this is revised replacing Warner's (1965) device by that of Horvitz et al.'s (1967) and Greenberg et al.'s (1969) as follows. A respondent in an SRSWR in n draws is requested to divulge one's true characteristic A/A^c saying "Yes"/"No" with a known probability $(1 - T)$ by applying Simmons's RRT choosing with probability p $(0 < p < 1)$ to say "Yes"/"No" truthfully about A and with probability $(1 - p)$ about B. Suppose θ_B is known and finding a total of n' "Yes" responses out of n responses an unbiased estimator of θ is then

$$\hat{\theta}_T = \frac{\left(\dfrac{n'}{n}\right) - (1 - T)(1 - p)\theta_B}{p + (1 - p)T}.$$

Its variance is

$$V(\hat{\theta}_T) = \frac{\phi_T (1 - \phi_T)}{n\left[T + (1 - T)p\right]^2},$$

writing $\phi_T = [T + (1 - T)p]\theta + (1 - T)(1 - p)\theta_B$. An unbiased estimator for

$$V\left(\hat{\theta}_T\right) \text{ is } v_T = \frac{\left(\dfrac{n'}{n}\right)\left(1 - \dfrac{n'}{n}\right)}{(n - 1)\left[p + (1 - p)T\right]^2}.$$

The authors have worked out appropriate combinations of p, T-values to achieve higher efficiency of $\hat{\theta}_T$ over $\hat{\theta}$. Chang and Liang (1996) have also extended their study to cover situations permitting partially untruthful responses to the RRT queries. Their algebra too is involved to merit recapitulation here.

Kajal Dihidar

Dihidar (2010a) applied Mangat and Singh's (1990) technique to Greenberg et al.'s (1969) unrelated response model when θ_B is unknown. The author considered DR cum RR from an SRSWR in n draws but following Chaudhuri's (2001) approach needed two independent "Yes"/"No" responses from each sampled person in the following way.

The first response with probability T $(0 < T < 1)$ is the true value y_i from the ith person or with probability $(1 - T)$ the response is I_i which is "Yes"/"No" about bearing A with probability p_1 or about B with probability $(1 - p_1)$. Similarly, an independent response is y_i with probability T and with probability $(1 - T)$ the response is J_i which is "Yes"/"No" about A with probability p_2 and "Yes"/"No" about the attribute B with probability $(1 - p_2)$. Here of course,

y_i = 1/0 if i bears A/A^c, 1/0 is "Yes"/"No"
x_i = 1/0 if i bears B/B^c, 1/0 is "Yes"/"No"
I_i = 1/0 if match/mismatch about A/B in the first and
J_i = 1/0 if match/mismatch about A/B in the second RR device. Writing
z_i = y_i with probability T
    = I_i with probability $(1 - T)$ and
Z_i' = y_i with probability T
    = J_i with probability $(1 - T)$, one gets

$$E_R(z_i) = Ty_i + (1 - T)\left[p_1 y_i + (1 - p_1)x_i\right],$$
$$E_R(z_i') = Ty_i + (1 - T)\left[p_2 y_i + (1 - p_2)x_i\right],$$

yielding

$$r_i = \frac{(1 - p_2)z_i - (1 - p_1)z_i'}{(p_1 - p_2)},$$

for which $E_R(r_i)$ and $V_R(r_i) = \phi(y_i - x_i)^2$ writing

$$\phi = \frac{(1 - T)(1 - p_1)(1 - p_2)}{(p_1 - p_2)^2}\left[(1 - p_2)(T + p_1(1 - T)) + (1 - p_1)(T + p_2(1 - T))\right],$$

taking $p_1 \neq p_2$.

Since $V_i = V_R(r_i) = E_R(r_i^2) - y_i = E_R r_i(r_i - 1)$, one may unbiasedly estimate V_i by $r_i(r_i - 1) = v_i$, say. As a consequence, for any sampling design an unbiased estimator for $Y = \Sigma_1^N y_i$ follows in terms of r_is in a sample and for the variance of this estimator of Y in terms of r_is and v_is. But Dihidar (2010a) presents the results only for an SRSWR in n draws. An advantage in the presentation is that it was possible to give a concrete condition for the reduction in variance in estimation consequent to an appropriate choice of T by applying Mangat et al.'s (1990) method. Dihidar (2010a) further employed "shrinkage estimation technique" for a further improvement in estimating θ on utilizing a prior guess about its possible value as a follow up to Thompson's (1968) basic theory for it was earlier pursued in the RR context by Singh et al. (2007). A detailed account here may be cumbersome however variance estimators are provided by Dihidar (2010a).

Dihidar (2010b) further applied Mangat and Singh's (1990) modification technique to Singh and Joarder's (1997) "unknown repeated trial model" discussed earlier and also on the "forced response model" of Boruch (1972) as also discussed by Chaudhuri and Mukerjee (1988) in the context of "Attributes" and further includes quantitative characteristics as well. The approach was first to apply Chaudhuri's (2001) method to first estimate y_i unbiasedly by an RR, r_i, related to a selected individual labeled i, work out the $V_R(r_i)$ and then derive an alternative to r_i as say, r'_i also unbiased for y_i, utilizing Mangat et al.'s (1990) technique of mingling a possible DR on y_i with an RR from i and show that $V_R(r'_i) < V_R(r_i)$ under conditions analogs to those applicable to Singh et al.'s (1990) modification on Warner's (1965) and Greenberg et al.'s (1969) unrelated model, in each case without stipulating how a sample of respondents is to be selected. Finally the author allows the sample to be chosen by a general unequal probability sampling. Naturally simulation-based numerical illustrations are presented to show the advantages, if any, through Mangat and Singh's (1990) technique applied on the corresponding classical RR techniques on which it is applied. Advantages are proclaimed just in case the estimated coefficients of variation show tendencies to decline on implementing Mangat and Singh's (1990) corrective measures.

Chang, Wang, and Huang and Huang

Chang et al.'s (2004) RR device needs two independent SRSWR in n_1 and n_2 draws, $n_1 + n_2 = n$. A person sampled is required to say "Yes" if the person bears the sensitive attribute A with an unknown probability T and if the person bears the complement A^c is required to say "Yes" meaning the person bears A with an assigned probability P_j and say "No" that is, the person bears A^c with probability $(1 - P_j)$, if the person is in the jth sample ($j = 1, 2$) .

Then, $\mu_j = T\theta + (1 - P_j)(1 - \theta)$ is the probability of a "Yes" response from a person chosen from the jth sample; θ being the proportion in the community bearing A. Writing $\bar{Z}_j$ as the proportion of persons in the jth sample saying "Yes", they derive and propose

$$\hat{\theta} = \frac{\left(\bar{Z}_1 - \bar{Z}_2\right) + \left(P_1 - P_2\right)}{P_1 - P_2}, \quad P_1 \neq P_2$$

and

$$\hat{T} = \frac{(1 - P_2)\bar{Z}_1 - (1 - P_1)\bar{Z}_2}{(\bar{Z}_1 - \bar{Z}_2) + (P_1 - P_2)}$$

as unbiased estimators for θ and a biased estimator respectively for T. They have worked out

$$V(\hat{\theta}) = \frac{1}{(P_1 - P_2)^2}\left[\frac{\mu_1(1 - \mu_1)}{n_1} + \frac{\mu_2(1 - \mu_2)}{n_2}\right]$$

which admits an unbiased estimator

$$v(\hat{\theta}) = \frac{1}{(P_1 - P_2)^2}\left[\frac{\bar{Z}_1(1 - \bar{Z}_1)}{n_1 - 1} + \frac{\bar{Z}_2(1 - \bar{Z}_2)}{n_2 - 1}\right]$$

They, however, observed that $\hat{T}$ is biased for T.
  They also calculated

$$\text{Bias}(\hat{T}) = \frac{1}{(P_1 - P_2)^2}\frac{1}{\theta^2}\left[\frac{(T - 1 + P_2)\mu_1(1 - \mu_1)}{n_1} + \frac{(T - 1 + P_1)\mu_2(1 - \mu_2)}{n_2}\right]$$

and gave a formula for

$$MSE(\hat{T}) = E(\hat{T} - T)^2$$

$$= \frac{1}{(P_1 - P_2)^2\theta^2}\left[\frac{(T - 1 + P_2)^2\mu_1(1 - \mu_1)}{n_1} + \frac{(T - 1 + P_2)^2\mu_2(1 - \mu_2)}{n_2}\right]$$

of which respectively their proposed estimators are

$$\hat{\text{Bias}}(\hat{T}) = \frac{1}{(P_1 - P_2)^2}\frac{1}{(\hat{\theta})^2}\left[\frac{(\hat{T} - 1 + P_2)}{n_1}\hat{Z}_1(1 - \hat{Z}_1) + \frac{(\hat{T} - 1 + P_1)}{n_2}\hat{Z}_2(1 - \hat{Z}_2)\right]$$

$$\hat{MSE}(\hat{T}) = \frac{1}{(P_1 - P_2)^2}\frac{1}{(\hat{\theta})^2}\left[\frac{(\hat{T} - 1 + P_2)^2\hat{Z}_1(1 - \hat{Z}_1)}{n_1} + \frac{(\hat{T} - 1 + P_1)^2\hat{Z}_2(1 - \hat{Z}_2)}{n_2}\right].$$

They compared Mangat's (1994) estimator of θ and $\hat{\theta}$ in terms of the respective MSEs and presented numerical results concerning their relative performances.

Huang (2004) considers a dichotomous population with an unknown proportion θ bearing a stigmatizing attribute A and an SRSWR in n draws and supposing a person bearing A to agree to say so has a probability T while those bearing A^c are willing to tell the truth if asked for. In a DR survey to estimate θ his estimator is $\hat{\theta}_D$, the proportion of "Yes" responses with a bias $B(\hat{\theta}_D) = E(\hat{\theta}_D) - \theta$ and an MSE as

$$MSE(\hat{\theta}_D) = \frac{\theta T(1 - \theta T)}{n} + \theta^2(1 - T)^2.$$

He suggests no estimators for these. In going for RRTs he first addresses the problem of estimating this T along with that of θ as well.

For a DR, the probability for a "Yes" response is

$$\lambda_1 = \theta T.$$

For RR, the same is

$$\lambda_2 = \theta p(1 - T) + (1 - p)(1 - \theta)$$

because if the respondent bears A and draws the card with an instruction to say so, then the "Yes" answers will follow with probability $(1 - T)$ and if the respondent draws the complementary card the respondent will straightaway say "Yes" if the respondent does not bear A. Consequently,

$$\hat{\theta} = \frac{p\hat{\theta}_1 + \hat{\theta}_2 - (1 - p)}{2p - 1} \quad \text{and} \quad \hat{T} = \frac{(2p - 1)\hat{\theta}_1}{p\hat{\theta}_1 + \hat{\theta}_2 - (1 - p)}$$

are, respectively unbiased and biased estimators for θ, T, writing $\hat{\theta}_j$ as the observed proportions of "Yes" responses in the DR and RR respectively for $j = 1, 2$. The formula for variance is

$$V(\hat{\theta}) = \frac{\theta(1 - \theta)}{n} + \frac{p(1 - p)(1 - \theta T)}{n(2p - 1)^2}.$$

It has been shown that this variance is less than the variance of Warner's unbiased estimator for θ. Also the unbiased estimator for $V(\hat{\theta})$ is

$$v(\hat{\theta}) = \frac{\hat{\theta}(1 - \hat{\theta})}{n - 1} + \frac{p(1 - p)}{(n - 1)(2p - 1)^2} - \frac{p(1 - p)\hat{\theta}_1}{n(2p - 1)^2}.$$

The appropriate form for the MSE of T is

$$MSE(\hat{T}) = \frac{T(1-T)}{n} + \frac{p(1-p)T^2(1-\theta T)}{n(2p-1)^2 \theta^2}.$$

The other results are not dealt with here because of the present reviewer's limited interest in developments derived from SRSWRs.

Kim and Elam and Kim and Warde

We already discussed how Mangat et al. (1997) followed by Singh et al. (2000) presented their protective amendments on the RRT of Moors (1971). The efforts of Kim and Warde (2004a) geared toward that direction.

They, as observed with most researchers in the field of RRTs, restricted their study to SRSWR in n draws. Each respondent is required to truthfully answer "Yes" or "No" about bearing an innocuous attribute B directly. Those responding "Yes" are requested to implement an RR device R_1. Under R_1, a person is to honestly report one's true attribute B or the negative of it with probability P_1 and about the sensitive attribute A or A^c with probability $1 - P_1$. Anyone reporting "No" about the direct query about B is requested to respond according to a second RR device R_2. Under R_2 a person is to truthfully report about bearing A with probability P and about A^c with probability $1 - P$, the RR being as usual "Yes" or "No" only. Out of the n DRs, let n_1 and $n_2 (= n - n_1)$ be the sampled numbers of "Yes" and "No" responses.

Writing θ, θ_B, respectively as the unknown proportions in the population bearing A, B respectively, the probability of getting a "Yes" response from the sampled respondents implementing the R_1 device is

$$P_Y = P_1\theta + (1 - P_1)\theta_B = P_1\theta + (1 - P_1)$$

Because those who try R_1 do bear B. Writing $\hat{Y}$ for the sampled proportion of "Yes" responses among those who try R_1

$$\hat{\theta}_a = \frac{\hat{Y} - (1 - P_1)}{P_1}$$

is an unbiased estimator for θ. Then,

$$V(\hat{\theta}_a) = \frac{P_Y(1 - P_Y)}{n_1 P_1^2} = \frac{(1-\theta)\left[P_1\theta + (1 - P_1)\right]}{n_1 P_1}.$$

Writing P_X as the probability of a "Yes" response out of those who try the R_2 device

$$P_X = P\theta + (1-P)(1-\theta) = (2P-1)\theta + (1-P)$$

Writing $\hat{X}$ as the observed sample proportion of "Yes" responses from those who try the R_2 device

$$\hat{\theta}_b = \frac{\hat{X} - (1-P)}{(2P-1)}$$

is an unbiased estimator for θ and

$$V(\hat{\theta}_b) = \frac{P_X(1-P_X)}{n_2 P_2} = \frac{\theta(1-\theta)}{n-n_1} + \frac{P(1-P)}{(n-n_1)(2P-1)^2}.$$

So they proposed

$$\hat{\theta}_m = \frac{n_1}{n}\hat{\theta}_a + \frac{n_2}{n}\hat{\theta}_b$$

as a pooled unbiased estimator for θ having the variance as

$$V(\hat{\theta}_m) = \frac{n_1}{n^2}\left[\frac{(1-\theta)\{P_1\theta + (1-P_1)\}}{P_1}\right] + \frac{n-n_1}{n}\left[\theta(1-\theta) + \frac{P(1-P)}{(2P-1)^2}\right]$$

This is described by Kim and Warde (2004a) in the "mixed randomized response" model. They rightly observe that (1) n_1 (and hence n_2 as well) is a random variable and that (2) a high value of the number of respondents to opt for R_1 should turn out for the sake of efficiency in estimation. They judiciously addressed the relevant issues in view of (1) and (2). Keeping the level of protection affordable by Warner's and Simmons's model on par, Lanke (1976) derived the following result, which is

$$P = \frac{1}{2} + \frac{P_1}{2P_1 + 4(1-P_1)\theta_1}$$

for every P_1 and every θ_1. Since $\theta_1 = 1$ in this case these authors observe the special case (2) which reduces as

$$P = \frac{1}{2} + \frac{P_1}{2P_1 + 4(1-P_1)_1} = \frac{1}{(2-P_1)}$$

Using this and writing $\lambda = m/n$ they get

$$\text{Var}(\hat{\theta}_m) = \frac{\theta(1-\theta)}{n} + \left[\frac{(1-P_1)\{\lambda P(1-\theta) + (1-\lambda)\}}{nP_1^2} \right].$$

This forms a basis for comparison between the mixed randomized response model and that of Moors's model discussed by Greenberg et al. (1971). Detailed empirical findings have been tabulated regarding such efficiency comparisons; variance estimation is not taken up by Kim and Warde (2004a).

Kim and Warde (2004b) were inspired by the works of Hong et al. (1994), who extended Warner's (1965) RRT by considering stratified simple random sampling. The latter restricted it to proportional sample allocations to the strata. But Kim and Warde (2004b) considered not only Warner's (1965) RRT but also RRT models given to us by Mangat and Singh (1990) and also Mangat's (1994) RRT treating stratified simple random sampling with replacement (Str. SRSWR) with Neyman's optimal allocation formulae. By dint of elaborate algebra, Kim and Warde (2004b) have clearly shown the advantages that accrue from stratified over nonstratified sampling with respect to all the three RR techniques. Kim and Warde (2004a) illustrate further the efficacy of Str. SRSWR with optimal allocation sampling rule over unstratified SRSWR when employed along with their mixed RR technique. In both these works their algebraic findings are supplemented by numerical findings through simulations. We are not keen to reproduce their algebra because of the present writer's apathy against SRSWR rather than general unequal probability sampling. Kim and Elam (2005) cover more of these Str. SRSWR using these four RRTs with the additional coverage of not only truthful RRs but also not so truthful RRs that seem more natural human instincts.

Kim, Tebbs, and An; Chua and Tsui: Their Works

The starting point of the development by Kim et al. (2006) is Mangat's (1994) modification of Warner's (1965) RRT—both already dealt with in great detail so far.

For θ, the proportion bearing a sensitive attribute in a given community, Mangat (1994) employs the estimator

$$\hat{\theta}_M = \frac{\hat{\lambda} - 1p}{p}.$$

Here $\hat{\lambda}$ is the proportion of "Yes" as the RR in an SRSWR in n draws and p is the probability of choosing a card needed to report "Yes" or "No" truthfully about bearing A rather than A^c, its complement. They have given its variance as

$$V(\hat{\theta}_M) = \left[\frac{\theta(1-\theta)}{n} + \frac{(1-\theta)(1-p)}{n_p} \right].$$

A modified estimator vis-à-vis that proposed for θ by Kim et al. (2006) based on the same set-up is

$$\theta_M^* = \begin{cases} 1-p & if \ \lambda \leq 1-p \\ \hat{\lambda} & if \ 1-p < \lambda < 1 \end{cases}.$$

Kim et al. (2006) have proposed a Bayes estimator for θ as follows on noting the likelihood for θ on observing t as the total number of "Yes" responses in Mangat's (1994) RRT from an SRSWR in n draws as $n - t \ T(\theta|t) = [p\theta + (1-p)]^t$ $[p(1-\theta)]^{n-t}$ and choosing a β prior for θ as

$$q(\theta) = \frac{1}{B(\alpha,\beta)} \theta^{\alpha-1}(1-\theta)^{\beta-1}, \quad \alpha > 0, \quad \beta > 0, \ (0 \leq \theta \leq 1).$$

With the usual approach of deriving the posterior expectations of θ permitting a square error loss for θ their Bayes estimator for θ is

$$\hat{\theta}_B = \frac{\displaystyle\sum_{j=0}^{t} \binom{t}{j} d^{t-j} B(\alpha + j + 1, n + \beta - t)}{\displaystyle\sum_{j=0}^{t} \binom{t}{j} d^{t-j} B(\alpha + j, n + \beta - t)}$$

writing $d = t(1-p)$. Then they compared the MSEs as

$$MSE(\hat{\theta}_B) = E_t\left[\left(\hat{\theta}_B - \theta\right)^2 \right] = \sum_{t=0}^{n} (\hat{\theta}_B - \theta)^2 \binom{n}{t} (p\theta + 1 - p)^t [p(1-\theta)]^{n-t};$$

similarly

$$MSE(\hat{\theta}_M) = E_t[(\hat{\theta}_M - \theta)^2] \ \text{and} \ MSE(\theta_M^*)$$

They made detailed empirical studies with numerical calculations. Further they extended the studies to stratified sampling in usual manners as partly covered in this text as well. Chua and Tsui (2000) have given us a novel RRT purported to induce a sense of privacy being inherently protected of a respondent on request to give out truthful facts about bearing a stigmatizing attribute A in an SRSWR chosen in n draws. Their RR scheme is applied in two stages with respect to each person sampled. In the first stage each person sampled is requested to randomly choose one number by SRSWOR with respect to the persons selected by SRSWR in n draws from a given set of integers $\Omega = \{1, ..., K, ..., m\}$, $m \geq n$. Without divulging the number say, K, so drawn to the investigator, in the second stage is requested to draw randomly m values from either a distribution F if bearing A or from a different distribution G if bearing A^c and report the Kth value in the increasing order. Both F and G are fully specified with known means μ_F, $\mu_G (\mu_F \neq \mu_G)$ and known variances σ_F^2, σ_G^2.

Equipped with these RRs denoted as $X_1, ..., X_i, ..., X_n$ so that

$$E(X_i) = \theta\mu_F + (1 - \theta)\mu_G$$

Chua and Tsui (2006) employ

$$\hat{\theta} = \frac{\bar{X} - \mu_G}{\mu_F - \mu_G}$$

as their unbiased estimator for θ, writing $\bar{X}$ as the sample mean of X_i's. Writing μ_{ij}, σ_{ij}^2 as the mean, variance of the ith order statistic in the m variables generated from the distribution J which is either F or G, they derive the variance as

$$V(\hat{\theta}) = \frac{1}{n(\mu_F - \mu_G)^2}\left[\theta\left(\sigma_F^2 + \mu_F^2 - \frac{1}{m}\sum_{i=1}^{m}\mu_{iF}^2\right)\right.$$

$$+ |(1 - \theta)\left(\sigma_G^2 + \mu_F^2 - \frac{1}{m}\sum_{i=1}^{m}\mu_{iF}^2\right) + \theta(1 - \theta)\frac{1}{m}\sum_{i=1}^{m}(\mu_{iF} - \mu_{iG})$$

$$+ \left. \frac{m - n}{m(m - 1)}\sum_{i=1}^{m}\left\{\theta\mu_{iF} + (1 - \theta)\mu_{iG} - \left(\theta\mu_F + (1 - \theta)\mu_G\right)\right\}^2\right].$$

They observe (1) $V(\hat{\theta})$ is minimum if $m = n$, (2) if $\lim_{m\to\alpha}(m - n)/(m - 1) = 1$,

$$\lim_{m\to\alpha}V(\hat{\theta}) = \frac{\theta(1 - \theta)}{n} + \frac{\theta\sigma_F^2 + (1 - \theta)\mu_G^2}{n(\mu_F - \mu_G)^2}$$

which is the variance of the unbiased estimator for θ given by Franklin (1989);

also (3) with $\mu_F = P$, $\mu_G = 1 - P$, $\sigma_F^2 = P(1 - P) = \sigma_G^2$, this

$$\lim_{m \to \alpha} V(\hat{\theta})$$

reduces to the variance of the estimator of θ given by Warner (1965). They have cited various special forms of F, G along with discussions spelling out efficiency levels numerically attained in varying situations.

Carlos N. Bouza

Bouza (2009) has a unique contribution to the RR literature to record the role of ranked set sampling (RSS) rivaling SRSWR in yielding serviceable estimation through quantitative RR data for a finite population mean of values relating to a stigmatizing real-valued feature. He considers the RRT on a sensitive quantitative variable described by Chaudhuri and Stenger (1992). According to it, for a finite population $U = (1, ..., i, ..., N)$ of a known size N the problem is to estimate $Y = \sum_{i=1}^{N} y_i$ for y_i being the stigmatizing value on a real variable y for the person labeled i in U. For this a sampled person is presented two boxes containing respectively cards marked $\underline{A} = (A_1, ..., A_T)$ and $\underline{B} = (B_1, ..., B_L)$ and is requested to choose randomly one card marked A_k, from $\underline{A}$ and return it on noting the value and independently choosing one valued B_u from $\underline{B}$ and return it to the lot by noting its value. The RR elicited from this person is then

$$Z_i = A_K y_i + B_u.$$

Then, $E_R(Z_i) = y_i \mu_A + \mu_B$ and $V_R(Z_i) = y_i^2 \sigma_A^2 + \sigma_B^2$ writing

$$\mu_A = \frac{1}{T} \sum_{K=1}^{T} A_K, \quad \mu_B = \frac{1}{L} \sum_{u=1}^{L} B_u,$$

$$\sigma_A^2 = \frac{1}{T} \sum_{1}^{T} (A_K - \mu_A)^2, \quad \sigma_B^2 = \frac{1}{L} \sum_{1}^{L} (B_u - \mu_B)^2.$$

Then, $R_i = (Z_i - \mu_B)/\mu_A$, provided $\mu_A \neq 0$, has $E_R(R_i) = y_i$ and $V_R(R_i) = [y_i^2 \sigma_A^2 + \sigma_B^2)/\sigma_A^2] = V_i$, say, obviously with $\sigma_A \neq 0$.

To estimate Y, Bouza (2009) generates the above R_i's from a sample of persons i taken, following the RSS method of McIntyre (1952) further developed and reviewed by Chen, Bai and Sinha (2004) among others.

RSS is implemented as follows. First m independent SRSWRs in m draws each are selected. Using hypothetical values of y_i or the values x_i of a variable x closely correlated with y, the units i of $U = (1, ... i, ..., N)$ are to be ranked in an increasing order. A sample s when drawn its units are supposed to be ranked. For the ith sample, $i = 1, ..., m$, the person is given the rank i ($i = 1, ..., m$), using all the m units in the particular sample, $s(i)$ at hand, $i = 1, \ldots, m$. The person ranked first in $s(1)$, second in $s(2)$, ..., m in $s(m)$ are then approached to generate the m RRs following the method described above. This procedure is independently repeated r times on taking $rm = n$, which is the sample size of the RSS. Each of these r, replicates of m samples is called a "Cycle." The RR from "the ith ranked sample" by the Chaudhuri and Mukerjee's RRT treated by Bouza (2009) for the tth cycle is

$$z_{it} = A_k y_{it} + B_u \text{, say.}$$

Then, $E_R(z_{it}) = \mu_A y_{it} + \mu_B$ and $r_{it} = (z_{it} - \mu_B)/\mu_A$ has $E_R(r_{it}) = y_{it}$.
Then,

$$\bar{r}_t = \frac{1}{m} \sum_{i=1}^{m} r_{it} \quad \text{has} \quad E_R(\bar{r}_t) = \frac{1}{m} \sum_{i=1}^{m} y_{it}$$

$$E_p E_R(\bar{r}_t) = E_p \left[\frac{1}{m} \sum_{i=1}^{m} y_{it} \right] = E_p(\bar{y}_t)$$

say, with

$$\bar{y}_t = \frac{1}{m} \sum_{i=1}^{m} y_{it}, \quad = \bar{Y}.$$

Then, $\bar{r}(RSS) = (1/r)(1/m) \sum_{t=1}^{r} \sum_{i=1}^{m} r_{it}$ is their proposed estimator for $\bar{Y}$. The theory presented seems too intricate. So, we refrain from further explanation.

Joe Kerkvliet

This is a follow-up of the treatment of logit modeling reported in Maddala's (1983). The stigmatizing use of cocaine by college students vis-à-vis academic performance and socioeconomic characteristics is the question under discussion by Kerkvliet (1994). He considers "forced response" RRT of Boruch (1972), Maddala (1983), and Chaudhuri and Mukerjee (1988). As usual he has no regards for any sampling scheme except SRSWR. His main concern is to examine the magnitudes of p_y, p_n which are the probabilities to be fixed for the respondents to say "Yes" or "No" irrespective of their bearing A or A^c so that $(1 - p_y - p_n)$ may be appropriate as the probability with which each may truthfully say "Yes" or "No" corresponding to the actual traits, being the sensitive A or its negation A^c. The sensitive trait treated by Kerkvliet (1994) is cocaine consumption by a college student. Academic performance and personal habits are endogenous deterministic factors. The magnitudes of p_y, p_n are determined by the sum of the last four digits in a student's Social Security number expressed as a fraction of $36 = 9 + 9 + 9 + 9$, treating the above sum as a random variable.

The probability that a student consumes cocaine, namely θ is to be estimated on postulating that for

$$\eta = \frac{\theta}{1 - \theta}, \quad \log_e \eta = \phi(\underline{X})$$

for a suitable function ϕ, say, linear, in terms of $\underline{X} = (X_1, X_2)$ with X_1, X_2 reflecting academic performance and social behavior.

So, he postulates the "logistic regression model"

$$L(\eta) = \frac{\pi}{i|y_i=1}\left[p_y + \left(1 - p_y - p_n\right)\frac{e^{\beta_1 X_{1i} + \beta_2 X_{2i}}}{1 + e^{\beta_1 X_{1i} + \beta_2 X_{2i}}} \right]$$

$$\times \frac{\pi}{i|y_i=0}\left[p_n + \left(1 - p_y - p_n\right)\frac{1}{1 + e^{\beta_1 X_{1i} + \beta_2 X_{2i}}} \right].$$

Estimation of β_1, β_2, and θ then follows and numerical values are tabulated based on live data.

N.J. Scheers

This work predominantly talks about the philosophy of RRT as an appropriate tool to protect a respondent from the risk of revealing incriminating characteristics like under reporting of tax liabilities, cheating, use of drugs and thus reduce propensities to tell lies or suppress truths and makes references to validation studies regarding RR-data versus DR-data. Most of the material covered in this work was already reviewed by Chaudhuri and Mukerjee (1988). Only the logistic regression modeling for example, covered by Kerkvliet and Scheers and Dayton (1988) was not discussed by Chaudhuri and Mukerjee (1988). But in our present work this point has been discussed. One glaring slip in the present work is the use of the multiplier in the variance estimators throughout, though in view of the very big sample sizes in the illustrations, this theoretical deficiency hardly affects the empirical results.

Umesh and Peterson

This addresses philosophical issues questioning the inherent utility of the RR approach. Success of RRT depends on the rate of truthful responses to questions about socially stigmatizing personal traits randomized rather than by direct inquiries. If DRs were trustworthily available in abundance RRs would be of little worth yielding less efficient estimates with inflated variances. Still RRT is employed believing it instills faith in respondents that their secrecy is protected inducing unrestricted true disclosures with mitigated chances of revelation. The unanswerable question is whether response bias is really eliminated or substantially curtailed. These aspects are succinctly discussed in this article which is a critical appraisal of the statistical developments, validational issues and substantive applications of RRTs in practice through the recounting of emerging growth in novel methods by ways of successive improvements over the preceding techniques.

As it is a review article most of its material is treated already by Fox and Tracy (1986) and by Chaudhuri and Mukerjee (1988). But a few, for example, the ones noted below deserve to be mentioned as source materials of possible interest to the readers. They are by Himmelfarb and Edgell (1980), Poole and Clayton (1982), Clickner and Iglewiez (1980), Reinmuth and Geurts (1975), Lamb and Stem (1978), Fox and Tracy (1984), Kraemer (1980), Edgell et al. (1986), Gunel (1985), Spurrier and Padgett (1980), Pitz (1980), O'Hagan (1987), and Tracy and Fox (1981) among others like Volicer and Volicer (1982), Volicer et al. (1983), Zdep et al. (1979), Brewer (1981), Akers et al. (1983) dealing essentially with how to contend with drug users by suitable RRTs; exclusively, SRSWR forms the backbone.

Chris Gjestvang and Sarjinder Singh

It claims to be an improvement over the RRTs by Himmelfarb and Edgell (1980) and by Eichorn and Hayre (1983). It uses two scrambling variables instead of one as in the former and generates an RR by combining the two with preassigned probabilities. They also narrate how optimal RR may also be accommodated. Yet the SRSWR is the lone sampling method that constitutes the basis of this work.

Landsheer, Heijden, and Gils

This is a crucial report on how an RRT may fare as a capable technique to extract reliable information on sensitive issues. This examines illustratively how in respect of involvement of social security fraud even the miscreants may be induced to give out truthful responses trusting that RRs may enable to hide the secrets with mitigated risks of revelation. They illustrate applications of forced response technique (cf. Fox and Tracy, 1986) and Kuk's (1990) RRT to show how an investigator's capability of instilling understanding of an RRT in a respondent with adequate trust that the respondent's privacy will be preserved in spite of a true revelation by dint of an RRT. In generating acceptability of the RR theory in the general community of scientists and practitioners this work plus its bibliography may turn out a landmark publication.

Heijden and Gils

This exercise is similar to the one by Kerkvliet (1990) discussed earlier, illustrating how the forced response model (cf. Fox and Tracy, 1986) or Kuk's

(1990) model for data generation by randomization on sensitive characteristics may be utilized in estimating parameters about such features utilizing by dint of logistic regression modeling now covariate observations as
well. Like Kerkvliet (1990) these authors as well refer to earlier works by
Maddala (1983) and Scheers and Dayton (1988). As usual they also restrict
to SRSWR alone.

Heiden, Gils, Bouts, and Hox

Citing a practical problem of social relevance, namely counting the recipi-
ents of social welfare benefits exercising fraudulent tricks, this paper
reports the relative efficacies of the four contesting procedures, namely (1)
face-to-face direct interviewing, (2) computer-assisted self interviewing
(CASI) and (3,4) two RRTs namely forced response procedure (cf. Fox and
Tracy 1986) and Kuk's (1990) procedure. Validation study and adjusting
logistic regression modeling on introducing two-or-three explanatory variables in case of RR-studies constitute essential ingredients of this chapter
which avoids rigorous statistical analysis consistently with the spirit of
coverage of this chapter.

Tracy and Mangat

This is a review of what followed since 1984 upto which the findings in the
field of RR were covered by Chaudhuri and Mukerjee (1988). Though most of
its contents are already reviewed in the present book so far, a few interesting
citations in Tracy and Mangat (1996) have not caught our attention yet.

Ahsanullah and Eichhorn (1988) added to Eichhorn and Hayre's (1983)
work utilizing the multiplier scramble variable. Singh and Kathuria (1992)
modified Simmons's approach introducing two innocuous features. They
also referred to Chaudhuri and Adhikary's (1990) use of PPS sampling
and its follow up by Bansal et al. (1994) treating multicharacter surveys
using RRs. They also mention Tracy and Mukhopadhyay (1994) using
ordered sampling and Arnab's (1990) point that commutativity in design
and model expectation operation held more generally than envisaged by
Chaudhuri and Adhikary (1990). They also corroborate Arnab's (1994)
observation that a "non-negative variance estimator in a DR survey yields
one for the one derived from it for an RR survey though the converse is not
guaranteed."

Tracy and Mangat (1996) also mention the works of Little (1993), Ibrahim
(1990), and Bourke and Moran (1998), who apply model-based likelihood
theory, EM algorithm technique and multiple imputation approach to tackle
sensitive issues using predictive distributions. The present reviewer lacks
acumen to seriously deal with these procedures. They also mention the case
of testing if a respondent is lying and then studying RRs in estimation as
treated by Krishnamoorthy and Raghavarao (1993) and by Lakshmi and

Raghavarao (1992). They also referred to Warner's (1986) telephone-based RRs and Bourke's (1990) EM-based MLEs. Further they covered Ljungqvist's (1993) approach for respondent jeopardy as follow ups of the works of Lanke and Flinger et al. (1977). They also refer to the empirical RR studies by Danermark and Swensson (1987) and by Duffy and Waterton (1988) concerning smoked hashish or marijuana and also cocaine use study by Kerkvliet (1994) as we mentioned earlier.

Sanghamitra Pal

This combines an RR technique with the item count technique (ICT) permitting an option to a respondent to go for either RRT or ICT without divulging to the interviewer the option actually exercised. The respondent considers estimating for a finite population $U = (1, ..., i, ..., N)$ the mean $\theta_A = (1/N)\sum_1^N \theta_i$, where

$$\theta_j = 1 \quad \text{if } i \text{ bears a sensitive attribute } A$$

$$= 0 \quad \text{if the person bears the complement } A^c.$$

The ICT is applied on taking two independent samples s_1 and s_2 on following the same sampling design p with positive inclusion probabilities π_i for i and π_{ij} for $i, j(i \neq j)$ in U. A person i in s_1 is given two lists with G items common in both but in one list the $(G + 1)$th item implying either A^c or F^c or both A^c and F^c simultaneously. The proportion θ_F in the population bearing F is supposed to be known.

A person in s_1 labeled i, say, is to give out y_i which means the number of features in the first list applicable to the person. A person labeled, say, j is asked to report the number x_j of the traits in the second list applicable to the person. For this ICT given by Chaudhuri and Christofides their unbiased estimator for θ_A is

$$\hat{\theta}_A = \frac{1}{N}\sum_{i \in s_1}\frac{y_i}{\pi_i} - \frac{1}{N}\sum_{j \in s_2}\frac{x_j}{\pi_j} + (1 - \theta_F)$$

with its unbiased variance estimator as

$$v\left(\hat{\theta}_A\right) = \left[\sum_{i<i'}\sum_{\in s_1}\left(\frac{\pi_i\pi_{i'} - \pi_{ii'}}{\pi_{ii'}}\right)\left(\frac{y_i}{\pi_i} - \frac{y_{i'}}{\pi_{i'}}\right)^2 + \sum\frac{y_i^2}{\pi_i^2}\alpha_i\right.$$

$$\left.\times \sum_{j<j'}\sum_{\in s_2}\left(\frac{\pi_j\pi_{j'} - \pi_{jj'}}{\pi_{jj'}}\right)\left(\frac{x_j}{\pi_j} - \frac{x_{j'}}{\pi_{j'}}\right)^2 + \sum\frac{x_j^2}{\pi_j^2}\alpha_j\right]\Bigg/N^2$$

Here

$$\alpha_i = 1 + \frac{1}{\pi_i} \sum_{i' \neq i} \pi_{ii'} - \sum_{i=1}^{N} \pi_i$$

$$\alpha_j = 1 + \frac{1}{\pi_j} \sum_{j' \neq j} \pi_{jj'} - \sum_{1}^{N} \pi_j.$$

In contrast with this ICT, the RRT recommended by Pal (2007) works as follows. Out of the integers 1, 2, ..., K, ..., G, the ith person in the sample s_1 is requested to choose one, K at random, denoting this random realization as u_i and the requested RR is $I_i = \theta_i + u_i$. Then, $r_i = I_i - E(u_i) = I_i - (G + 1)/2$ has $E_R(r_i) = \theta_i$ with $V_R(r_i) = V_i$, known in terms of G admitting an unbiased estimator v_i for itself. Then, $\tilde{\theta} = (1/N) \sum_{i \in s_i} r_i/\pi_i$ estimates θ_A unbiasedly along with its unbiased variance estimator as

$$v(\tilde{\theta}) = \frac{1}{N^2} \left[\sum_{i<} \sum_{i'} \left(\frac{\pi_i \pi_{i'} - \pi_{ii'}}{\pi_{ii'}} \right) \left(\frac{r_i}{\pi_i} - \frac{r_{i'}}{\pi_{i'}} \right)^2 + \sum_{i \in s_1} \frac{r_i^2}{\pi_i^2} \alpha_i + \sum_{i \in s_1} \frac{v_i}{\pi_i} \right]$$

At this stage, Pal (2009) introduces a novelty in approach by permitting a sampled person i in s_1 to decide with an unknown probability $(1 - C_i)$ $(0 < C_i < 1)$ to yield the value y_i on employing Chaudhuri and Christofides's (2008) ICT or with probability C_i the value r_i by the RRT as above. Thus, the RR is

$$z_i = \theta_i + u_i = I_i \text{ with probability } C_i$$

$$= y_i \text{ with probability } (1 - C_i).$$

Again, another independent realization from the same person i in s_1 is g_i with the same distribution as of z_i and independently of it. Then the author recommends

$$t_i = \frac{1}{2}(z_i + g_i)$$

for which

$$E_R(t_i) = E_R(z_i) = E_R(g_i)$$

and

$$w_i = \frac{1}{4}(z_i - g_i)^2 \quad \text{has} \quad E_R(w_i) = V_R(z_i) = V_R(g_i).$$

Each sampled person j in s_2 is on request to give with probability $(1 - C_j)$ the value $1 + u_j - F$ following with the ICT or with probability C_j the RR as x_j.

Then, similarly a second response is independently made by j in s_2 by the same method. Calling the responses as z'_j, g_j corresponding to t_i, g_i respectively, it follows that

$$t'_j = \frac{1}{2}\left(\frac{z'_j + g'_j}{2}\right)$$

has $E_R(g'_j) = E_R(z'_j) = E_R(t'_j)$ and $1/4(z'_j - g'_j)^2$ is an unbiased estimator w'_j for $V_R(z'_j) = V_R(g'_j)$. Letting

$$e_1 = \frac{1}{N}\sum_{i \in s_1}\frac{t_i}{\pi_i}, \quad e_2 = \frac{1}{N}\sum_{j \in s_2}\frac{t'_j}{\pi_j}, \quad \text{and} \quad \theta_F = \frac{1}{N}\sum_{j=1}^{N}F_j.$$

Pal derives $\theta^* = e_1 - e_2 + 1 - \theta_F$ as an unbiased estimator for θ_A with an unbiased variance estimator for θ^* as

$$v\left(\theta^*\right) = \sum_{i<i' \in s_1}\sum\left(\frac{\pi\pi_{i'} - \pi_{ii'}}{\pi_{ii'}}\right)\left(\frac{t_i}{\pi_i} - \frac{t_{i'}}{\pi_{i'}}\right)^2 + \sum_{i \in s_1}\frac{w_i}{\pi}\alpha_i$$

$$+ \sum_{j<j' \in s_2}\sum\left(\frac{\pi_j\pi_{j'} - \pi_{jj'}}{\pi_{jj'}}\right)\left(\frac{t'_j}{\pi_j} - \frac{t'_{j'}}{\pi_{j'}}\right)^2 + \sum_{j \in s_2}\frac{w'_j}{\pi_j}$$

as an unbiased estimator for $V(\theta^*)$.

Amitava Saha

Dalenius and Vitale's (1974) RRT-based results derived under SRSWR has been expanded by Chaudhuri (2002) as described in Chapter 2 of the present book. Saha (2003) applied the approach of Chaudhuri and Mukerjee's (1985, 1988) ORR procedure of splitting a given sample into two parts: one giving the DRs and the other the RRs, so as to improve upon the efficiency in estimation through Rao–Blackwellization. Saha (2003) illustrated numerically with fictitious data how Dalenius and Vitale (1974)-based DR-cum-RR data by ORR shows improvement over CRR approach for SRSWR and more prominently for SRSWOR and still more so for Rao–Hartley–Cochran (1962)-based sample data. We omit the details in this chapter.

Strachan, King, and Sarjinder Singh

They consider estimation for the regression model

$$\underline{y} = X\beta + \underline{e}$$

when observations on the vector y of the dependent variables are not directly available as in the case of stigmatizing situations but instead "scrambling" variables either in the additive form of Himmelfarb and Edgell (1980) or in the productive form treated by Eichhorn and Hayre (1989) are used as distorters with known distributions with known parameters. Sarjinder Singh et al. (1996) provide basic materials for their initialization of the relevant developments. They consider method of moments (MM), maximum likelihood (MLE), and Bayesian estimation methods. They need application of EM (expectation maximization) algorithm and Markov Chain Monte Carlo (MCMC) techniques involving Gibbs samplers and Metropolis–Hastings algorithms (MHA) drawing upon the complicated Bayesian calculation procedures developed by Gelfand and Smith's (1990), Tanner's (1993) and Smith and Robert's (1993) works. This book is unable to deal with further details.

Sanghamitra Pal

The implicit RRT of Takahasi and Sakasegawa (1977) presented based on SRSWR in Chaudhuri and Mukerjee (1988) and extended by Pal (2007) to cover general sampling schemes has been described under "Takahasi and Sakasegawa's Scheme Modified by Pal (2007)" of Chapter 3 of this book. In a subsequent work, Pal (2009) has further extended the work to include the unique ORR coverage of this IRR (Implicit RR) technique of Takahasi et al. (1983) following twin approaches of Chaudhuri and Saha (2005), Chaudhuri and Dihidar (2009), and Pal (2009) respectively when a respondent (1) openly gives out the truth about bearing A or A^c and (2) alternatively, either opts for a DR or this specific IRR without divulging the option actually exercised.

To treat case (1), let us recall from Chapter 3 under "Takahasi and Sakasegawa's Scheme Modified by Pal (2007)" that three independent samples s_j ($j = 1, 2, 3$) are drawn by the same sampling scheme and three independent IRRs r_{ji} ($j = 1, 2, 3$) are derived with say, $E_R(r_{ji}) = M_{ji}$ and $\bar{M}_j = 1/N \sum_{i=1}^{N} M_{ji}$ for $j = 1,\ 2,\ 3$, leading to the unbiased estimator for $\theta_A = 1/N \sum_{i=1}^{N} y_i$ as $\hat{\theta}_A = \hat{\theta} - 3$ with

$$\hat{\theta} = \sum_{j=1}^{3} \hat{\theta}_j, \quad \hat{\theta}_j = \frac{1}{N} \sum_{i \in s} b_{sji}\, r_{ji}, \quad j = 1, 2, 3.$$

If the respondents in the part A_j of s_j give out the DRs while those in $B_j = s_j - A_j$ give out the IRRs by Takahasi and Sakasegawa (1983) procedure, then Pal (2009) revises $\hat{\theta}_j$ by

$$\tilde{\theta}_j = \frac{1}{N} \left[\sum_{i \in A_j} b_{sji} M_{ji} + \sum_{i \in B_j} b_{sji} r_{ji} \right]$$

and hence $\tilde{\theta} = \sum_{j=1} \tilde{\theta}_j$ and $\tilde{\theta}_A = \tilde{\theta} - 3$ as an unbiased estimator for θ_A. Following the approach of Chaudhuri and Saha (2004) it could (a) be shown that $V(\tilde{\theta}_A) < V(\hat{\theta}_A)$ and (b) find an unbiased estimator for $V(\tilde{\theta}_A)$. To cover the case (2) Pal first notes that the third sample s_3 is not needed. Only s_1 and independent of it s_2 are required. Keeping the notations in tact for the elements related to IRR concerning s_1 and s_2 what Pal needs additionally is to presume every sampled person i with an unascertainable probability C_i $(0 < C_i < 1)$ is supposed to give out the DR about A or with probability $(1 - C_i)$ give out the IRR. With this background, Pal derives an unbiased estimator for θ_A along with an unbiased variance estimator essentially along the direction followed by Chaudhuri and Dihidar (2009). Details need not be reproduced further.

Ardo van den Hout and P.G.M. van der Heijden

These authors recognize RR set-up as a particular case of "statistical disclosure control and misclassification."

For example, for Warner's (1965) RRT based on SRSWR one has the structure:

$$\underline{\lambda} = \begin{pmatrix} \lambda = Prob(\text{Yes}) \\ 1 - \lambda = Prob(\text{No}) \end{pmatrix} = \begin{pmatrix} p & 1-p \\ 1-p & p \end{pmatrix} \begin{pmatrix} \theta \\ 1-\theta \end{pmatrix}$$

$$= \underline{P}\ \underline{\theta}, \text{ say.}$$

Here $\underline{P}$ is the known matrix for probabilities for misclassification known to both the inquirer and the respondent.

They note that when MM is applied to estimate $\underline{\theta}$, the estimators may not be feasible for certain device of $\underline{P}$. But the method of MLE is not responsive to this possibility, the estimation demanding a study of the "boundary problem" if an estimated value falls in the "boundary" of the "parameter space." They also recognize the need for the application of the EM algorithm when solution of the likelihood equation (LE) turns out difficult to come by. Relevant mathematics are omitted to avoid complications unwarranted in this book.

D.E. Stem and R.K. Steinhorst

These authors describe how RR may be serviceably generated by self-administered and mailed questionnaire or by telephone interview methods. Detailed practical methods are given with illustrated results. As the present reviewer lacks experience in such exercises it seems prudent not to discuss further.

Ravindra Singh, Sarjinder Singh, and N.S. Mangat

These authors formulated a detailed theory for a mailed survey technique to procure data on stigmatizing characteristics. The algebra is very absorbing as it involves SRSWOR rather than SRSWR and also stratification of the respondents into potential respondents in the first attempt and the dichotomous complementary set of the people. They also refer to the similar earlier works by Singh et al. (1993).

D.R. Bellhouse

Although the present reviewer failed to notice a convincing reason why one needs it, Bellhouse addresses the problem of estimating the finite population correlation coefficient between two variables, both deemed stigmatizing. He considered apt, not clearly spelled out why so, to imply three specific RRTs namely (1) Greenberg et al.'s (1969) unrelated response technique, (2) Pollok and Bek's (1976) additive scrambler and (3) their multiplicative scrambler to ensure anonymity in the true response. Most importantly, he considered general probability sampling designs illustrating in particular SRSWOR in n draws. Specifically, he considered "Random Permutation Model" first to derive suitably optimal estimators for the variances of x and of y and their covariance deriving therefrom a transform to get the estimated correlation coefficient. The details are cumbersome and hence omitted. The finite population correlation coefficient between x and y is namely,

$$\rho_N = \frac{N\sum_1^N y_i x_i - \left(\sum_1^N y_i\right)\left(\sum_1^N x_i\right)}{\sqrt{N\sum y_i^2 - \left(\sum y_i\right)^2}\sqrt{N\sum x_i^2 - \left(\sum x_i\right)^2}}$$

and may be estimated from a DR survey by

$$r = \frac{\sum_{i\in s} 1/\pi_i\left(y_i - \sum_{i\in s} y_i/\pi_i\right)\left(x_i - \sum_{i\in s} x_i/\pi_i\right)}{\sqrt{\sum_{i\in s} 1/\pi_i\left(y_i - \sum_{i\in s} y_i/\pi_i\right)^2}\sqrt{\sum_{i\in s} 1/\pi_i\left(x_i - \sum_{i\in s} x_i/\pi_i\right)^2}}.$$

Efforts are needed to develop refinements to switch over to the utilization of procedures when only RR survey data are available. No literature of relevance seems available yet. We refrain from including details of Bellhouse's (1995) studies on estimating ρ_N.

Yan, Zaizai and Nie, Zankan

Conceding that the notations used earlier remain unchanged, without further clarifications, let us note the two conceptual definitions below as quoted from Yan and Nie's (2004). The entire development is restricted to SRSWR in n draws only.

Definition 9.1

For the RR design with "Yes" (Y)–"No" (N) responses, only

$$\Delta = |Pr(A|Y) - Pr(A|N)|$$

is the degree of protection.

Definition 9.2

A design d_1 is better than a design d_2 if

$$\Delta_{d1} = \Delta_{d2} \quad \text{and} \quad V_{d_1}(\hat{\theta}_A) < V_{d_2}(\hat{\theta}_A) \ \forall \ \theta_A \in [0,1]$$

Nayak's (1994) definitions are different.

Yan and Nie (2004) insist on comparing any two RR procedures only through the above Definitions 9.1 and 9.2.

Recalling Warner's (1965) RRT with $\lambda_A = p_w \theta_A + (1 - p_w)(1 - \theta_A)$,

$$\hat{\theta}_{Aw} = \frac{\left[\dfrac{nA}{n} - (1 - p_w) \right]}{(2p_w - 1)}, \quad p_w \neq \frac{1}{2}.$$

$$V\left(\hat{\theta}_{Aw}\right) = \frac{\lambda_w (1 - \lambda_w)}{n(2p_w - 1)2},$$

they get

$$\Delta_w = \frac{(2p_w - 1)}{\lambda_w (1 - \lambda_w)} \theta_A (1 - \theta_A).$$

Treating the Simmons's URL with known θ_B and writing

$$\lambda_s = p_s \theta_A + (1 - p_s)\theta_B, \quad \hat{\theta}_{As} = \frac{(n_A/n) - (1 - p_s)\theta_B}{p_s}, \quad V(\hat{\theta}_{As}) = \frac{\lambda_s(1 - \lambda_s)}{np_s^2},$$

one has

$$\Delta_s = \frac{p_s}{\lambda_s(1 - \lambda_s)}\theta_A(1 - \theta_A),$$

Hence they derive

$$\frac{\lambda_w(1 - \lambda_w)}{2p_{w-1}} = \frac{\lambda_s(1 - \lambda_s)}{p_s},$$

as the condition for Warner's and Simmons's RRTs to be equally protective of secrecy.

Hence they have the

Theorem 9.17

Provided $p_w > 1/2$,

1. If $p_w > \dfrac{(1 + p_s)}{2}$, Warner's RRT is better than that of Simmons.

2. If $p_w < \dfrac{1 + p_s}{2}$, Warner's is worse.

3. If $p_w = \dfrac{1 + p_s}{2}$, the two RRTs are on a par.

Yan and Nei by the same criterion also compared other RRTs as well.

Epilogue

Since the literature on RRTs and other ICTs have already grown in volume and is still growing, it seems an updated review as the one presented here is due.

Without being apologetic, the present author has clearly expressed his views about RR theory through his works covered in the first eight chapters, while in this chapter he has presented in a nutshell the activities of his contemporaries with great respect. Several aspects of these deserve further development with major overhauling. His latest plan is to endeavor to achieve as much as possible toward this end, possibly in a new book or in published papers until that target is achieved.

10

Illustrative Simulated Empirical Findings

Warner's RR versus Unrelated Question Model-Based RR in Unequal Probability Sampling as against Equal Probability Sampling

Let us artificially consider $N = 113$ particular persons. For each of them labeled $i = 1, ..., N$ (=113), let y_i denote that the ith person is "a habitual gambler" when it is valued 1 and "not so" when it is valued 0; likewise, let $x_i = 1$ if ith person prefers cricket to football and $x_i = 0$ in the opposite case. Let the individual (y_i, x_i) values be known resulting in

$$\theta = \frac{\sum_1^N y_i}{N} = 0.8230 \quad \text{and} \quad \overline{X} = \frac{\sum_1^N x_i}{N} = 0.7345$$

Suppose a sample of size n is considered using (1) SRSWR, (2) SRSWOR, and (3) by Rao–Hartley–Cochran Scheme separately and independently from $U = (1, ..., N = 113)$. We consider (1) the CRR case when each sampled person gives an RR by a "method" specified and the (2) case when the first n_1 (<n) of them gives an RR and the remaining $n_2 = n - n_1$ of them give a DR each. We illustrate the situation when $n = 33$, $n_1 = 24$, and $n_2 = 9$. The, RR schemes considered separately are (A) Warner's and (B) Simmons' with prescribed p and (p_1, p_2), respectively. To draw an RHC sample, the person's last month pocket expenses in dollars z_i are taken as the known size measures.

Since we treat this hypothetical example with known N, (y_i, x_i, z_i) and specified n, n_1, and n_2 we may implement the actual sample selection by SRSWR, SRSWOR, and RHC scheme and CRR or ORR data gathering independently any number of times we illustrate the case with R, the number of replications being 1000.

Then, for θ we may employ the usual estimator $\hat{\theta}$ with a variance estimator υ generically for CRR and ORR and for these three sampling schemes specified as above. Then, considering the standardized pivotal

$$\delta = \frac{\hat{\theta} - \theta}{\sqrt{\upsilon}},$$

we may construct the 95% confidence interval (CI) as $(\hat{\theta} - 1.96\sqrt{v}, \hat{\theta} + 1.96\sqrt{v})$. Then, we may find the (1) ACP, as the percentage of replicates for which the calculated CI covers θ, (2) the ACV, as the average of $100\ \sqrt{v}/\hat{\theta}$ over these 1000 replicates, and also the (3) AL, as the average value over these 1000 replicates of the width of the CI, which is $2 \times 1.96\sqrt{v}$. The findings are tabulated below. In RHC sampling, the group sizes are taken as $N_i = [N/n]$ or $[N/n] + 1$ subject to $\Sigma_n N_i = N$.

Briefly, the RHC scheme may be narrated as follows. To take a sample of n units from a population $U = (1, ..., i, ..., N)$ of size N with z_i as positive size measures with $Z = \Sigma_{zi}$ and $P_i = z_i/Z$, $i \in U$ first by successive SRSWORs of sizes N_i from U are taken to construct n disjoint groups of N_i units for $i = 1, ...,$ n with $\Sigma_n N_i = N$ writing Σ_n to denote the sum over the n groups. Then, independently across the groups, one unit is chosen with probability proportional to P_i from within each group. Writing Q_i as the sum of the P_i's in the ith group, N_i in number, RHC's unbiased estimator for $Y = \Sigma_l^N y_i$ is

$$ t_{RHC} = \sum_n y_i \frac{Q_i}{P_i}, $$

writing y_i for the y value of the unit chosen from the ith group.

An unbiased estimator for $\theta = Y/N$ is then $\hat{\theta} = t_{RHC}/N$ and an unbiased estimator for the variance of $\hat{\theta}$ is $v = \Sigma_n \Sigma_n Q_i Q_i (y_i/P_i - y_i/P_i)^2/N^2$, writing $\Sigma_n \Sigma_n$ as the sum over the distinct pairs of n_{c2} groups avoiding repetition.

When SRSWR or SRSWOR is used $\hat{\theta}$ is the sample mean with a well-known variance estimator formula generically for v.

TABLE 10.1

A Comparative Illustration (cf. Chapter 5)

A. Warner's Method

	CRR			ORR		
p	*ACV*	*ACP*	*AL*	*ACV*	*ACP*	*AL*
SRSWR						
(0.30)	32.3	97.0	0.94	27.8	98.2	0.34
(0.66)	37.4	97.2	1.15	35.8	97.0	1.03
SRSWOR						
(0.30)	27.0	93.7	0.81	22.7	73.3	5.70
(0.66)	35.9	92.4	1.04	35.5	95.4	0.90
RHC						
(0.30)	42.8	97.5	1.22	38.5	96.5	1.14
(0.66)	54.9	97.4	1.48	51.7	96.5	1.37

TABLE 10.1   (continued)

A Comparative Illustration (cf. Chapter 5)

B. Simmons' Unrelated (URL) Question Model

(p_1, p_2)	CRR			ORR		
	ACV	*ACP*	*AL*	*ACV*	*ACP*	*AL*
SRSWR						
(0.29, 0.95)	8.5	94.6	0.27	8.0	92.6	0.25
(0.28, 0.97)	8.4	94.0	0.26	8.1	93.2	0.25
SRSWOR						
(0.17, 0.98)	7.0	91.3	0.22	6.9	91.9	0.22
(0.09, 0.96)	7.0	92.3	0.22	6.8	94.0	0.22
RHC						
(0.04, 0.97)	11.1	91.5	0.37	10.9	90.5	0.37
(0.25, 0.97)	11.1	93.7	0.37	10.8	91.4	0.37

Note: Intuitively seems okay.

TABLE 10.2

Privacy Protected vis-à-vis Efficiency Achieved (cf. Chapter 6)

Warner's Model

	L_i	$L_i(1)$	V_i	$L_i(1)$
p	0.4	0.5		
0.44	0.343	0.440	17.111	1.029
0.59	0.489	0.590	7.466	1.066

Forced Response

		L_i		$L_i(1)$		V_i		$\bar{J}_i$
p_1	p_2	0.4		0.5		$y_i = 1$	$y_i = 0$	
0.62	0.54	0.330		0.425		9.70	9.20	1.08
0.88	0.32	0.340		0.435		5.44	2.64	1.72

Mangat and Singh

		L_i	0.4	0.5	V_i	$\bar{J}_i$
p	T		$L_i(1)$			
0.44	0.52		0.644	0.731	0.92	1.54
0.34	0.69		0.721	0.795	0.47	2.07

Note: Strengthens intuition.

TABLE 10.3

Privacy Protected vis-à-vis Efficiency Achieved Based on Survey Data (cf. Chapter 6)

		Warner, SRSWR			
p		ACV	ACP	AL	$\bar{J}_i$
0.58		166.77	98.20	2.376	1.053
0.39		82.66	96.50	1.786	1.102
		Warner, SRSWOR			
p					
0.64		54.29	95.1	1.19	1.102
0.89		74.97	94.9	1.48	1.170
		Warner, RHC			
p					
0.60		54.54	97.2	2.406	1.083
0.30		42.79	97.5	1.218	1.381
		Forced Response, SRSWR			
p_1	p_2				
0.55	0.88	13.04	52.8	0.486	1.087
0.38	0.89	17.78	56.0	0.694	0.862
		Forced Response, SRSWOR			
p_1	p_2				
0.55	0.88	10.71	51.9	0.408	1.087
0.36	0.88	15.75	51.8	0.647	0.854
		Forced Response, RHC			
p_1	p_2				
0.55	0.88	30.02	98.9	1.098	1.087
0.38	0.89	26.44	36.5	1.469	0.862
		Simmons' URL, SRSWR			
p_1	p_2				
0.58	0.19	22.54	99.30	0.702	1.517
0.37	0.22	20.47	98.80	0.629	2.785
		Simmons' URL, SRSWOR			
p_1	p_2				
0.58	0.83	7.88	94.40	0.249	2.286
0.37	0.46	11.33	97.50	0.356	2.285
		Simmons' URL, RHC scheme			
0.58	0.92	11.85	92.30	0.394	4.597
0.40	0.91	11.98	93.30	0.405	8.401

Note: Forced response technique works competitively only when combined with the RHC scheme.

With this background, we selectively quote from the unpublished PhD thesis by A. Saha (2006).

Simulated Illustration of a Numerical Study of "Protection of Privacy" in RR Surveys

Quoting from the works of Chaudhuri et al. (2009) on "Protection of Privacy" let us present the following:

> We consider a community of 117 households (HH) constituting a population $U = (1, ..., i, ..., 117)$ bearing known size measures z_i as the last month's per capita consumption expenditure. Our parameter for estimation is θ = proportion of persons deliberately evading income tax liabilities. Let us write
>
> $y_i = 1$ if the principal earner of the ith HH evades IT
> $\quad = 0,$ otherwise;
> $x_i = 1$ if this ith person prefers football to cricket
> $\quad = 0,$ if otherwise, $I \in U$.
>
> Let $n = 34$ be the size of a sample proposed to be drawn by SRSWR, SRSWOR, or by the RHC technique. Let $\hat{\theta}$, v be defined as in Section 10.1, and so be the CI, ACP, AL, and ACV with 1000 as the number of replicates R. We shall use Warner's RRT, forced response RRT, Mangat and Singh's (1990), and Simmons' URL techniques. The usual notations including L_i as prior probability that ith person reveals the truth and the jeopardy values J_i's and the averages $\bar{J}_i$ are discussed in Chapter 6. Also, let V_i be the variance of an unbiased estimator r_i for y_i, $i \in U$.

Concluding Remarks

An interested reader is invited to have a look at the source materials of Chaudhuri et al. (2009) for further reading.

We may sum up asserting that quite a high level of statistical background is needed to assimilate most of the contents of this book.

References

Abul-Ela, A.A., Greenberg, B.G., and Horvitz, D.G. 1967. A multiproportion randomized response model. *JASA*, 62, 990–1008.

Adhikary, A.K., Chaudhuri, A., and Vijayan, K. 1984. Optimum sampling strategies for RR trials. *Int. Stat. Rev.* 52, 115–125.

Ahsanullah, M. and Eichhorn, B.H. 1988. On estimation of response from scrambled quantitative data. *Pak. J. Statist.* 4A, 83–91.

Akers, R.I., Massey, J., Clarke, W., and Lauer, R.M. 1983. Are self reports of adoloscent deviance valid? Biochemical measures, randomized response and the bogus pipeline in smoking behaviour. *Social Forces* 62, 234–251.

Anderson, H. 1975a. *Efficiency versus Protection in the RR Designs for Estimating Proportions.* Technical Report 9, University of Lund, Lund, Sweden.

Anderson, H. 1975b. *Efficiency versus Protection in a General RR Model.* Technical Report 10, University of Lund, Lund, Sweden.

Anderson, H. 1975c. *Efficiency versus Protection in RR Designs.* Mimeo notes. University of Lund, Lund, Sweden.

Anderson, H. 1976. Estimation of proportion through RR. *Int. Stat. Rev.* 44, 213–217.

Arnab, R. 1990. On commutativity of design and model expectations in randomized response surveys. *Comm. Stat. Theo. Math.* 19, 3751–3757.

Arnab, R. 1994. Non-negative variance estimator in randomized response surveys. *Comm. Stat. Theo. Math.* 23, 1743–1752.

Arnab, R. 1995a. On admissibility and optimality of sampling strategies in randomized response surveys. *Sankhya, Ser B.* 37(3), 385–390.

Arnab, R. 1995b. Optimal estimation of a finite population total under randomized response surveys. *Statistics* 27(1), 175–180.

Arnab, R. 1996. Randomized response trials: a unified approach for qualitative data. *Comm. Stat. Theo. Math.* 25(6), 1173–1183.

Arnab, R. 1998. Randomized response surveys: Optimum estimation of a finite population total. *Statistical Papers* 39(4), 405–408.

Arnab, R. 1999. On use of distinct respondents in RR surveys. *Biom. J.* 41(4), 507–513.

Arnab, R. 2000. Analysis of randomized response survey data. *Perspectives in Statistical Sciences.* Oxford University Press, pp. 18–26.

Arnab, R. 2001. Optimum sampling strategies under randomized surveys. *Stat. Papers* 54, 159–177.

Arnab, R. 2004. Optional randomized response techniques for complex designs. *Biom. J.* 46, 114–124.

Asok, C. 1980. A note on the comparison between simple mean and mean based on distinct units in sampling with replacement. *Am. Statist.* 34, 158.

Bansal, M.L., Singh, S., and Singh, R. 1994. Multi-character survey using randomized response technique. *Comm. Statist.—Theory Methods* 23(6), 1705–1715.

Basu, D. 1958. On sampling with and without replacement. *Sankhya* 20, 287–294.

Basu, D. 1969. Role of the sufficiency and likelihood principle in sample survey theory. *Sankhya A* 31, 441–454.

Basu, D. 1971. An essay on the logical foundations of survey sampling, Part I. In *Foundation of Statistical Inference* (FSI). Eds. Godambe, V.P. and Sprott, D.A., Holt, Rinehart and Winston, Toronto, Canada, 203–242.

Basu, D. and Ghosh, J.K. 1967. Sufficient statistics in sampling from a finite universe. *Bull. Int. Statist Inst.* 36, 850–859.

Bellhouse, D.R. 1995. Estimation of correlation in randomized response. *Survey Methodology* 21(1), 13–19.

Bhargava, M. and Singh, R. 1999. On the relative efficiency of certain randomized response strategies. *JISAS* 52(2), 245–253.

Bolfarine, H. and Zacks, S. 1992. *Predictions Theory for Finite Populations.* Springer–Verlag, NY.

Boruch, R.F. 1972. Relations among statistical methods for assuring confidentiality of social research data. *Soc. Sci. Res.* 1, 403–414.

Bourke, P.D. 1978. Randomized response designs with symmetric response for multiproportion situations. *Statistics Tidsskrift* 16, 197–207.

Bourke, P.D. 1990. Estimating a distribution for each category of a sensitive variate. *Comm. Stat. Theo. Math.* 19, 3233–3241.

Bourke, P.D. and Moran, M.A. 1988. Estimating proportions from randomized data using the EM algorithm. *J. Amer. Stat. Assoc.* 83, 964–968.

Bouza, C.N. 2009. Ranked set sampling and randomized response procedures for estimating the mean of a sensitive quantitative character. *Metrika*, DOI. 10.1007/s00/84 – 008 – 0191 – 6, 267–277.

Brewer, K.R.W. 1963. Ratio estimation and finite populations: Some results deducible from the assumption of an underlying stochastic process. *Aust. J. Stat.* 5, 93–105.

Brewer, K.R.W. 1979. A class of robust sampling designs for large-scale surveys. *JASA* 74, 911–915.

Brewer, K.R.W. 1981. Estimating marijuana usage using randomized response: Some paradoxical findings. *Aust. J. Stat.* 23, 139–148.

Cassel, C.M., Sarndal, C.E., and Wretman, J.H. 1976. Some result on generalized difference estimation and generalized regression estimation for finite populations. *Biometrika* 63, 615–620.

Chang, H.J. and Huang, K.C. 2001. Estimation of proportion and sensitivity of a qualitative character. *Metrika* 53, 269–280.

Chang, H.J. and Liang, D.H. 1996. A two-stage unrelated randomized response procedure. *Aust. J. Stat.* 33(1), 43–51.

Chang, H.J, Wang, C.L., and Huang, K.C. 2004. On estimating the proportion of a qualitative sensitive character using randomized response sampling. *Quality and Quantity* 38, 675–680.

Chaudhuri, A. 1987. Randomize response surveys of finite populations: A unified approach with quantitative data. *J. Stat. Plan. Inf.* 15, 157–165.

Chaudhuri, A. 1992. Randomized Response: Estimating mean square errors of linear estimators and finding optimal unbiased strategies. *Metrika* 39, 341–357.

Chaudhuri, A. 1993. Mean square error estimation in randomized response surveys. *Pak. J. Stat.* 9(2), 101–104.

Chaudhuri, A. 2000. Network and adaptive sampling with unequal probabilities. *CSAB* 50, 237–253.

Chaudhuri, A. 2001a. Using randomized response from a complex survey to estimate a sensitive proportion in a dichotomous finite population. *J. Stat. Plan. Inf.* 94, 37–42.

Chaudhuri, A. 2001b. Estimating sensitive proportions from unequal probability samples using randomized responses. *Pak. J. Stat.* 17(3), 259–270.

Chaudhuri, A. 2002. Estimating sensitive proportions from randomized responses in unequal probability sampling. *CSAB* 52, 315–322.

Chaudhuri, A. 2004. Christofides' randomized response technique in complex sample surveys. *Metrika* 60(3), 223–228.

Chaudhuri, A. and Adhikary, A.K. 1981. On sampling strategies with RR trials and their properties and relative efficiencies. Tech. Ref. ASC/81/5, Indian Statistical Institute, Calcutta.

Chaudhuri, A. and Adhikary, A.K. 1989. A note on handling linear randomized response. *J. Stat. Plan. Inf.* 22, 263–264.

Chaudhuri, A. and Adhikary, A.K. 1990. Variance estimation with randomized response. *Comm. Statist. Theory Meth.* 19(3), 1119–1126.

Chaudhuri, A. and Arnab, R. 1979. On the relative efficiencies of sampling strategies under a survey population model. *Sankhya C* 44, 92–101.

Chaudhuri, A. and Christofides, T.C. 2007. Item count technique in estimating the proportion of people with a sensitive feature. *J. Stat. Plan. Inf.* 137, 589–593.

Chaudhuri, A. and Christofides, T.C. 2008. Indirect questioning: How to rival randomized response techniques. *Int. J. Pure and Appl. Math.* 43(2), 283–294.

Chaudhuri, A. and Dihidar, K. 2009. Estimating means of stigmatizing qualitative and quantitative variables from discretionary responses randomized or direct. *Sankhya B* 71, 123–136.

Chaudhuri, A. and Mukerjee, R. 1985. Optionally randomized responses techniques. *CSAB* 34, 225–229.

Chaudhuri, A. and Mukerjee, R. 1987. Randomized responses techniques: A review. *Statist. Neerlandica* 41, 27–44.

Chaudhuri, A. and Mukerjee, R. 1988. *Randomized Responses: Theory and Techniques.* Marcel Dekker, New York, NY.

Chaudhuri, A. and Pal, S. 2002a. Estimating proportions from unequal probability samples using randomized responses by Warner's and other devices. *J. Ind. Soc. Agri. Stat.* 55(2), 174–183.

Chaudhuri, A. and Pal, S. 2002b. On certain alternative mean square error estimators in complex survey sampling. *J. Stat. Plan. Inf.* 104, 363–375.

Chaudhuri, A. and Pal, S. 2003. On a version of cluster sampling and its practical use. *J. Stat. Plan. Inf.* 113(1), 25–34.

Chaudhuri, A. and Pal, S. 2008. Estimating sensitive proportions from Warner's randomized responses in alternative ways restricting to only distinct units sampled. *Metrika* 68, 147–156.

Chaudhuri, A. and Pal, S. 2008. Estimating sensitive proportions from Warner's randomized responses. *Metrika* 68, 147–156.

Chaudhuri, A. and Saha, A. 2004. Utilizing covariates by logistic regression modelling in improved estimation of population proportions bearing stigmatizing features through randomized responses in complex surveys. *J. Ind. Soc. Agri. Stat.* 58(2), 190–211.

Chaudhuri, A. and Saha, A. 2005a. On relative efficiencies of optional versus compulsory randomization i responses: A simulation-based numerical study covering three RR schemes. *Pak. J. Stat.* 21(1), 87–98.

Chaudhuri, A. and Saha, A. 2005b. Optional versus compulsory randomized response techniques in complex surveys. *J. Stat. Plan. Inf.* 135, 516–527.

Chaudhuri, A. and Stenger, H. 1992. Theory and Methods of Survey Sampling. Marcel Dekker, Inc. NY.

Chaudhuri, A. and Stenger, H. 2005. *Survey Sampling: Theory and Methods* (2nd Edn). New York, NY: Chapman & Hall.

Chaudhuri, A. and Vos, J.W.E. 1988. Unified theory and strategies of survey sampling. North Holland, Amsterdam.

Chaudhuri, A., Adhikary, A.K., and Dihidar, S. 2000. Mean square error estimator in multi-stage sampling. *Metrika* 52, 115–131.

Chaudhuri, A., Adhikary, A.K., and Maity, T. 1998. A note on non-negative mean squared error estimation of regression estimators in randomized response surveys. *Statistical Papers* 39(4), 409–415.

Chaudhuri, A., Bose, M., and Dihidar, K. 2009a. Estimating sensitive proportions by Warner's randomized response technique using multiple randomized responses from distinct persons sampled. Statistical Papers—online through DOI: 10.1007/s 00362-009-0210-3.

Chaudhuri, A., Bose, M., and Dihidar, K. 2009b. Estimation of a sensitive proportion by Warner's randomized response data through inverse sampling. Statistical Papers—online through DOI: 10.1007/s 00362-009-0234-8.

Chaudhuri, A., Bose, M., and Ghosh, J.K. 2004. An application of adaptive sampling to estimate highly localized population segments. *J. Stat. Plan. Inf.* 121, 175–189.

Chaudhuri, A., Christofides, T.C., and Saha, A. 2009. Protection of privacy in efficient application of randomized response techniques. *Statist. Meth. Appl.* 18, 389–418.

Chaudhuri, A., Maity, T., and Roy, D. 1996. A note on competing variance estimation in randomized response surveys. *Aust. J. Stat.* 38(1), 35–42.

Chen, Z., Bai, Z., and Sinha, B.K. 2004. Ranked set sampling: theory and applications. Lecture Notes in Stat # 176. Springer, NY.

Chikkagoudar, M.S. 1966. A note on inverse sampling with equal probabilities. *Sankhya A* 28, 93–96.

Christofides, T.C. 2003. A generalized randomized response technique. *Metrika* 57, 195–200.

Christofides, T.C. 2010. Comments on a method of comparison of randomized response technique. *J. Stat. Plan. Inf* 140, 574–575.

Chua, T.C. and Tsui, A.K. 2000. Procuring honest responses indirectly. *J. Stat. Plan. Inf.* 90, 107–116.

Clickner, R.P. and Iglewiez, B. 1980. Warner's randomized response technique: The two sensitive question case. *South African Stat. J.* 14, 77–86.

Dalenius, T. and Vitale, R.A. 1974. *A New RR Design for Estimating the Mean of a Distribution*. Technical Report 78. Brown University, Providence, RI.

Danermark, B. and Swensson, B. 1988. Measuring drug use among Swedish adolescents: Randomized response versus anonymous questionnaires. *J. Off. Stat.* 3, 439–448.

Des, R. and Khamis, S.H. 1958. Some remarks on sampling with replacement. *AMS* 39, 550–557.

Dihidar, K. 2010a. On shrinkage estimation procedure combining direct and randomized responses in unrelated question model. *J. Ind. Soc. Agri. Stat.* 63(3), 2009 283–286; (also online at www.isas.org.in)

Dihidar, K. 2010b. Modifying classical randomized response techniques with provision for true response. *Cal. Stat. Assoc. Bull.* (in press).

Droitcour, E.M., Larson, E.J., and Schueren, F.J. 2001. The three card method: Estimating sensitive items with permanent anonymity of response. *Proc. Soc. Stat. Sec. ASA.*, Alexandria, VA.

Droitcour, J.A. and Larson, E.M. 2001. The Three Card Method: Estimating sensitive survey items—with permanent anonymity of response. *Proc. Ann. Meeting ASA* 245–250.

Droitcour, J.A. and Larson, E.M. 2002. An innovative technique for asking sensitive questions: The three card method. *Sociol. Methods Bull.* 75, 5–23.

Droitcour, J.A., Caspar, R.A., Hubbard, M.L., Parsley, T.L., Visseher, W., and Ezzati, T.M. 1991. The item count technique as a method of indirect questioning: A review of its development and a case study application. In *Measurement Error in Surveys.* Eds. Biemer, P.P., Groves, R.M., Lyburg, L.E., Mathiowetz, N., and Sudmar. S., John Wiley, NY.

Duffy, J.C. and Waterton, J.J. 1988. Randomized response vs direct questioning: Estimating the prevalence of alcohol related problems in a field survey. *Aust. J. Stat.* 30, 1–14.

Edgell, S.E., Himmelfarb, S., and Cira, D.J. 1986. Statistical efficiency of using two quantitative randomized response techniques to estimate correlation. *Psychological Bull.* 100, 251–256.

Edgell, S.E., Himmefarb, S., and Dunkan, K.L. 1982. Validity of forced response in a randomized response model. *Sociol. Methods Res.* 11(1), 89–100.

Eichhorn, B.H. and Hayre, L.S. 1983. Scrambled randomized response methods for obtaining sensitive quantitative data. *J. Statist. Plan. Inf.* 7, 307–316.

Eriksson, S.C. 1973. A new model for RR. *Int. Stat. Rev.* 41, 101–113.

Eriksson, S.C. 1980. Randomized Response, Unpublished Ph.D. thesis, Gothenburg University.

Flinger, M.A., Policello, G.E. and Sing, J. 1977. A comparison of two RR survey methods with consideration for the level of respondent protection. *Commun. Statist. Theory Methods* 6, 1511–1524.

Folsom, R.E. 1974. A randomized response validation study: Comparison of direct and randomized response reporting in DUI arrests. *Res. Triangle Inst. Rep.* No. 254–807.

Fox, J.A. and Tracy, P.E. 1980. The randomized response approach: Applicability to criminal justice research and evaluation. *Eval. Rev.* 4(5), 601–622.

Fox, J.A. and Tracy, P.E. 1984. Measuring associations with randomized response. *Soc. Sc. Res.* 13, 188–197.

Fox, J.A. and Tracy, P.E. 1986. *Randomized Response: A Method for Sensitive Surveys.* London: Sage.

Franklin, L.A. 1989a. Randomized response sampling from dichotomous population with continuous randomization. *Surv Methodol.* 15(2), 225–235.

Franklin, L.A. 1989b. A comparison of estimators for randomized response sampling with continuous distributions from dichotomous population. *Commun. Statist. Theory Method* 18(2), 489–505.

Gelfand, A.E. and Smith, A.F.M. 1990. Sampling-based approaches to calculating marginal densities. *JASA* 85, 398–409.

Geurts, M.D. 1980. Using a randomized response design to eliminate non-response and response biases in business research. *J. Acad. Market. Sci.* 8(2), 83–91.

Godambe, V.P. 1955. A unified theory of sampling from finite populations. *JRSS* B 17, 269–278.

Godambe, V.P. 1970. Foundations of survey sampling. *Amer. Stat.* 24, 33–38.

Godambe, V.P. 1980. Estimation in RR trials. *Int. Statist. Rev.* 48, 29–32.

Godambe, V.P. and Joshi, V.M. 1965. Admissibility and Bayes estimation in sampling finite populations. *AMS* 36, 1707–1722.

Godambe, V.P. and Thompson, M.E. 1973. Estimation in sampling theory with exchangeable prior distributions. *AS* 1, 1212–1221.

Godambe, V.P. and Thompson, M.E. 1977. Robust near optimal estimation in survey practice. *BISI* 47(3), 129–146.

Greenberg, B.G., Abul-Ela, A.-L., Simmons, W.R., and Horvitz, D.G. 1969. The unrelated question RR model: Theoretical framework. *JASA* 64, 520–539.

Greenberg, B.G., Kuebler, R.R., Abernathy, J.R., and Horvitz, D.G. 1971. Application of the randomized response technique in obtaining quantitative data. *JASA* 66, 243–250.

Guerriero, M. and Sandri, M.F. 2007. A note on the comparison of some randomized response procedures. *J. Stat. Plan. Inf.* 137, 2184–2190.

Gunel, F. 1985. A Bayesian comparison of randomized and voluntary response sampling models. *Commun. Statist. Theory Method* 14, 2411–2435.

Gupta, S., Gupta, B., and Singh, S. 2002. Estimation of sensitivity level of personal interview survey questions. *J. Stat. Plan. Inf.* 100, 239–247.

Hanurav, T.V. 1966. Some aspect of unified sampling theory. *Sankhya* A 28, 175–206.

Hartley, H.O. and Rao, J.N.K. 1968. A new estimation theory for sample surveys. *Biometrika* 55, 547–557.

Hartley, H.O. and Rao, J.N.K. 1971a. Foundations of survey sampling (a Don Quixote tragedy). *Amer. Stat.* 25, 21–27.

Hartley, H.O. and Rao, J.N.K. 1971b. Some comments on labels: A rejoinder to a section of Godambe's paper, 'A reply to my critics.' *Sankhya, Ser. C* 37, 163–170.

Ha'jek, J. 1959. Optimum strategy and other problems in probability samplings. *Casopis Pet. Mat.* 84, 389–473.

Ha'jek, J. 1971. Comments on a paper by Basu, D. 1971, 203–242.

Hedayat, A.S. and Sinha, B.K., 1979. Randomized response: A data-gathering tool for sensitive characteristics. *Design and Inference Infinite Population Sampling.* Eds. Hedayat, A.S. and Sinha, B.K., New York, NY: John Wiley.

Hege, V.S. 1965. Sampling designs which admit uniformly minimum unbiased estimators. *CSAB* 14, 160–162.

Himmelfarb, S. and Edgell, S.E. 1980. Additive constant model: A randomized response technique for eliminating evasiveness to quantitative response questions. *J. Psychol. Bull.* 87, 525–530.

Horvitz, D.G. and Thompson, D.J. 1952. A generalization of sampling without replacement from finite universe. *JASA* 47, 663–685.

Horvitz, D.G., Shah, B.V., and Simmons, W.R. 1967. The unrelated question RR model. *Proc. Social Statist. Sect. ASA*, 65–72.

Huang, K.C. 2004. A survey technique for estimating the proportion and sensitivity in a dichotomous finite population. *Stat. Neerlandice.* 58, 75–82.

Hubbard, M.L., Casper, R.A., and Lessler, J.T. 1989. Respondent reactions to item count lists and randomized response. *Proc. Social Statist. Sect. ASA*, 544–548.

Ibrahim, J.G. 1990. Incomplete data in generalized linear models. *J. Amer. Stat. Assoc.* 85, 765–769.

Kerkvliet, J. 1994. Estimating a logit model with randomized data: The case of cocaine use. *Aust. J. Stat.* 36(1), 9–20.

Kim, A.Y.C. 1990. Asking sensitive questions indirectly. *Biometrika* 77, 436–438.

Kim, J.M. and Elam, M.E. 2005. A two-stage stratified Warner's randomized response model using optimal allocation. *Metrika* 61, 1–7.

Kim, J. and Warde, W.D. 2004a. A stratified Warner's randomized response model. *J. Stat. Plan. Inf.* 120(1–2), 155–165.

Kim, J. and Warde, W.D. 2004b. A mixed randomized response model. *J. Stat. Plan. Inf.* 120, 155–165.

Kim, J.M., Tebbs, J., and An, S.W. 2006. Extensions of Mangat's randomized response model. *J. Stat. Plan. Inf.* 136(4), 1554–1567.

Korwar, R.M. and Serfling, R.J. 1970. On averaging over distinct units in sampling with replacement. *AMS* 41(6), 2132–2134.

Kraemer, H.C. 1980. Estimation and testing of bivariate association using data generated by the randomized response technique. *J. Psychol. Bull.* 87, 304–308.

Krishnamoorthy, K. and Raghavarao, D. 1993. Untruthful answering in repeated randomized response procedures. *Can. J. Stat.* 21, 233–236.

Kuk, A.Y.C. 1990. Asking sensitive questions indirectly. *Biometrika* 77(2), 436–438.

Lahiri, D.B. 1951. A method of sample selection providing unbiased ratio estimators. *BISI* 33(2), 133–140.

Lakshmi, D.V. and Raghavarao, D. 1992. A test for detecting untruthful answering in randomized response procedures. *J. Stat. Plan. Inf.* 31, 387–390.

Lamb, C. and Stem, D.E., Jr 1978. An empirical validation of the randomized response technique. *J. Market. Res.* 15, 616–621.

Landsheer, J.A., Heijden, P.V.D., and Gils, G.V. 1999. Trust and understanding, two psychological aspect of randomized response. *Quality Quantity* 33, 1–12.

Lanke, J. 1975a. Some contribution to the theory of survey sampling. PhD Thesis, University of Lund, Lund, Sweden.

Lanke, J. 1975b. On the choice of unrelated question in Simmons' version of RR. *JASA* 70, 80–83.

Lanke, J. 1976. On the degree of protection in randomized interviews. *Int. Stat. Rev.* 44, 197–203.

Leysieffer, R.W. 1975. Respondent jeopardy in RR procedures. Technical Report M338. Department of Statistics, Florida State University, Tallahassee, Florida.

Leysieffer, R.W. and Warner, S.L. 1976. Respondent jeopardy and optimal design in RR models. *JASA* 71, 649–656.

Little, R.J.A. 1982. Models for non-response in sample surveys. *J. Amer. Stat. Assoc.* 77, 237–261.

Liu, P.T. and Chow, L.P. 1976. A new discrete quantitative RR model. *JASA* 71, 72–73.

Liu, P.T., Chow, L.P., and Mosley, W.H. 1975. Use of RR technique with a new randomizing device. *JASA* 70, 329–332.

Ljungqvist, L. 1993. A unified approach to measures of privacy in randomized response models: a utilitarian perspective. *JASA* 88, 97–103.

Loynes, R.M. 1976. A symptotically optimal RR procedures. *J. Amer. Stat. Assoc.* 71, 924–928.

Maddala, G.S. 1983. *Limited Dependent and Qualitative Variables in Econometrics.* Cambridge University Press, New York.

Mangat, N.S. 1992. Two stage randomized response sampling procedure using unrelated question. *JISAS* 44(1), 82–88.

Mangat, N.S. 1994. An improved randomized response strategy. *JRSS, Ser. B* 56(1), 93–95.

Mangat, N.S. and Singh, R. 1990. An alternative randomized response procedure. *Biometrika* 77(2), 439–442.

Mangat, N.S. and Singh, R. 1991. An alternative approach to randomized response survey. *Statistica* 51(3), 327–332.

Mangat, N.S. and Singh, S. 1994. Optional randomized response model. *JISAS* 32, 71–75.

Mangat, N.S., Singh, R., and Singh, S. 1992. An improved unrelated question randomized response strategies. *CSAB* 42, 277–281.

Mangat, N.S., Singh, R., Singh, S., and Singh, B. 1993. On Moors' randomized response model. *Biom. J.* 35, 727–32.

Mangat, N.S., Singh, R., Singh, S., Bellhouse, D.R., and Kashani, H.B. 1995. On efficiency of estimator using distinct respondents in randomized response survey. *Surv. Methodol.* 21, 21–23.

McYntire, G.A. 1952. A method of unbiased selection sampling using ranked sets. *Aust. J. Agri. Res.* 3, 385–390.

Midzuno, H. 1952. On the sampling system with probabilities proportionate to survey of sizes. *AISM* 3, 99–107.

Miller, J.D. 1985. The nominative technique: A new method of estimating heroin prevalence. *NIDA Research Monograph* 57, 104–124.

Miller, J.D. 2001. The nominative technique: A new method of estimating heroin prevalence. 104–124.

Miller, J.D., Cisin, I.H., and Harrel, A.V. 1986. A new technique for surveying deviant behavior: Item count estimates of marijuana, cocaine and heroin. ASA paper, St. Petersberg, FL.

Moors, J.J.A. 1971. Optimization of the unrelated question RR model. *JASA* 66, 627–629.

Mukerjee, R. and Sengupta, S. 1989. Optimal estimation of finite population total under a general correlated model. *Biometrika* 76, 789–794.

Nathan, G. 1988. A bibliography on randomized responses: 1965–1987. *Survey Methodol.* 4(2), 331–346.

Nayak, T.K. 1994. On randomized response surveys for estimating a proportion. *Comm. Statist. Theory Method,* 23(3), 3303–3321.

Nayak, T.K. and Adeshiyan, S.A. 2009. A unified framework for analysis and comparison of randomized response survey of binary characteristics. *J. Stat. Plan. Inf.* 139, 2757–2766.

Ohlsson, E. 1989. Variance estimation in the Rao–Hartley–Cochran procedure. *Sankhya, Ser B* 51, 348–361.

O'Hagan, A. 1987. Bayes linear estimators for randomized response models. *JASA* 82, 580–585.

Padmawar, V.R. and Vijayan, K. 2000. Randomized response revisited. *J. Stat. Plan. Inf.* 90, 293–304.

Pal, S. 2002. Contributions to emerging techniques in survey sampling. Unpublished Ph.D. thesis, Indian Statistical Institute, Kolkata, India.

Pal, S. 2007a. Extending Takahashi–Sakasegawa's indirect response technique to cover sensitive surveys in unequal probability sampling. *CSAB* 59, 265–276.

Pal, S. 2007b. Estimating the proportion of people bearing a sensitive issue with an option to item count lists and randomized response. *Statist. Trans.* 8(2), 301–310.

Pal, S. 2008. Unbiasedly estimating the total of a stigmatizing variable from a complex survey on permitting options for direct or randomized responses. *Metrika* 49, 157–164.

Pal, S. 2009. Extending Takahasi–Sakasegawa's indirect response technique to cover sensitive surveys in unequal probability sampling permitting direct answers. Unpublished.

Pathak, P.K. 1962. On simple random sampling with replacement. *Sankhya* A 24, 287–302.

Pitz, G. 1980. Bayesian analysis of randomized response models. *J. Psychol. Bull.* 87, 209–212.

Pollock, K.H. and Bek, Y. 1976. A comparison of three RR models for quantitative data *JASA* 71, 884–886.

Poole, W.K. and Clayton, A.C. 1982. Generalization of a contamination model for continuous type random variables. *Commun. Statist. Theo. Methods* 11, 1733–1742.

Quatember, A. 2009. To copy with nonresponse and untruthful answering. Different questioning designs for different variables. In *Proceedings of the NTTS 2009—Conferences on New Techniques and Technologies for Statistics*, www.ntts2009.eu.

Rao, C.R. 1952. Some theorems on minimum variance estimation. *Sankhya* 12, 27–42.

Rao, J.N.K. 1979. On deriving mean square errors and other nonnegative unbiased estimators in finite population sampling. *JISA* 17, 125–136.

Rao, J.N.K. and Bellhouse, D.R. 1978. Optimal estimation of a finite population mean under generalized random permutation models. *J. Stat. Plan. Inf.* 2, 125–141.

Rao, J.N.K., Hartley, H.O., and Cochran, W.G. (RHC). 1962. On the simple procedure of unequal probability sampling without replacement. *JRSS* B 24, 482–491.

Reinmuth, J.E. and Geurts, M.D. 1975. The collection of sensitive information using a two stage randomized response model. *J. Market. Res.* 12, 402–407.

Royall, R.M. 1970. On finite population sampling theory under certain linear regression models. *Biometrika* 57, 377–387.

Saha, A. 2003. On efficacies of Dalenius–Vitale technique with compulsory versus optional randomized responses from complex surveys. *CSAB* 21(2).

Saha, A. 2004. On efficacies of Dalenius–Vitale technique with compulsory versus optional randomized responses from complex surveys. *CSAB* 54, 223–230.

Saha, A. 2006. Developing theories and inference procedures for practical survey problems. Unpublished PhD thesis, Indian Statistical Institute, Kolkata, India.

Scheers, N.J. 1992. A review of randomized response techniques in measurement and evaluation in counseling and development, *Meas. Eval. Couns. & Dev.* 25, 27–41.

Scheers, N.J. and Dayton, C.M. 1988. Covariate randomized response models. *JASA* 83, 969–974.

Sen, A.R. 1953. On the estimator of the variance in sampling with varying probabilities. *JISAS* 5(2), 119–127.

Sengupta, S. and Kundu, D. 1989. Estimation of finite population mean in randomized response surveys. *J. Stat. Plan. Inf.* 23, 117–125.

Singh, H.P. and Mathur, N. 2003. Modified optional randomized response sampling techniques. *JISAS* 56(2), 199–206.

Singh, J. 1976. A note on RR techniques. *Proc. ASA Soc. Stat. Sec.* 772.

Singh, J. 1978. A note on maximum likelihood estimation from randomized response models. *Proc. ASA Soc. Stat. Sec.*, 282–283.

Singh, R., Mangat, N.S., and Singh, S. 1993. A mail survey design for sensitive character without using randomization device. *Commun. Statist. Theory Method* 22(9), 2661–2668.

Singh, R., Singh, S., and Mangat, N.S. 2001. Mail survey design for sensitive quantitative variable. Unpublished.

Singh, S. and Joarder, A.H. 1997. Unknown repeated trials in randomized response sampling. *JISAS* 50, 70–74.

Singh, S. and Singh, R. 1992. Improved Franklin's model for randomized response sampling. *J. Ind. Stat. Assoc.* 30, 109–122.

Singh, S. and Singh, R. 1993. Generalized Franklin's model for randomized response sampling. *Commun. Statist. Theory Method* 22, 741–755.

Singh, S., Horn, S., and Chowdhuri, S. 1998. Estimation of stigmatized characteristics of a hidden gang in finite population. *Aust. NZ. J. Stat.* 4, 291–297.

Singh, S., Horn, S., Singh, R., and Mangat, N.S. 2003. On the use of modified randomization device for estimating the prevalence of a sensitive attribute. *Stat. in Transition* 6(4), 515–522.

Singh, S., Joarder, A.H., and King, M.L. 1996. Regression analysis using scrambled responses. *Aust. J. Stat.* 38(2), 201–211.

Singh, S., Mahmud, M., and Tracy, D.S. 2001. Estimation of mean and variance of stigmatized quantitative variables using distinct units in randomized response sampling. *Statist. Papers* 42, 403–411.

Singh, S., Singh, R., and Mangat, N.S. 2000. Some alternative strategies to Moor's model. *JASA* 66, 627–629.

Sinha, A. 2004. On efficacies of Dalenius–Vitale technique with compulsory versus optional randomized response from complex surveys. *CSAB* 54, 223–230.

Sinha, B.K. and Hedayat, A.S. 1991. Randomized response: a data-gathering tool for sensitive characteristics. In: *Design and Inference finite Population Sampling*. Eds. Hedayat, A. S. and Sinha, B. K., New York, NY: John Wiley & Sons, Chapter 11, 310–340.

Smith, A.F.M. and Roberts, G.O. 1993. Bayesian complication via the Gibbs sampler and related Markov chain Monte Carlo method (with discussion). *JRSS B* 55, 3–23, 53–102.

Spurrier, J. and Padgett, W. 1980. The application of Bayesian techniques in randomized response. *Social. Methods* 11, 533–544.

Stem, Jr., D.E. and Steinhorst, R.K. 1984. Telephone interview and mail questionnaire application of the randomized response model. *JASA* 79, 555–564.

Strachan, R., King, M., and Singh, S. 1998. Likelihood-based estimation of the regression model with scrambled responses. *Aust. NZ. J. Stat.* 40(3), 279–290.

Takahasi, K. and Sakasegawa, H. 1977. An RR technique without use of any randomizing device. *Ann. Inst. Stat. Math* 29, 1–8.

Tanner, M.A. 1993. *Tools for Statistical Inference: Methods for the Exploration of Posterior Distributions and Likelihood Functions*. 2nd Edn. New York, NY: Springer–Verlag.

Thionet, P. 1967. Application des numbres de sterling de 2e espece a un problem de sondage. *Revve de Statistique Appliquee* 15(4), 15–46.

Tracy, D. and Mangat, N.S. 1996. Some development in randomized response sampling during the last decades: A follow up of a review by Chaudhuri and Mukherjee. *JASS* 4(2/3), 147–158.

Tracy, D.S. and Mukhopadhyay, P. 1994. On UMVU-estimation under randomized response models. *Statistics* 25, 173–175.

Tracy, D.S. and Osahan, S.S. 1994. Random non-response on study variable versus on study as well as auxiliary variable. *Statistica* 54, 163–168.

Tracy, P. and Fox, J. 1981. The validity of randomized response for sensitive measurements. *Am. Sociol. Rev.* 46, 187–200.

Turner, F.C. 1998. Adolescent sexual behavior, drug use and violence: Increased reporting with computer survey technology. *Science* 280, 867–873.

Umesh, U.N. and Petersen, R.A. 1991. A critical evaluation of the randomized response method. *Soc. Meth. Res.* 20(1), 104–138.

Van der Heijden, P.G.M. and Van Gils, G. 1996. Some logistic regression models for randomized response data. *Proc. 11th Int. workshop on Stat. Modeling*, Orvieto, Italy, July 1996, 341–348.

Van der Heijden, P.G.M., Van Gils, G., Bouts, J., and Hox, J.J. 2000. A comparison of randomized response, computer-assisted self-interview and face-to-face direct questioning. Eliciting sensitive information in the context of welfare and unemployment benefit. *Soc. Meth. & Res.* 28, 505–537.

Van der Hout, A. and Van der Heijden, P.G.M., 2002. Randomized response, statistical disclosure control and misclassification: A review. *ISR* 30(2), 269–288.

Volicer, B.J. and Volicer, I. 1982. Randomized response technique for estimating alcohol use and noncompliance in hypertensives. *J. Stud. Alcohol* 43, 739–750.

Warner, S.L. 1965. RR: a survey technique for eliminating evasive answer bias. *JASA* 60, 63–69.

Wiseman, F. 1975–76. Estimating public opinion with the randomized response model. *Public Opin. Quart.* 39, 507–513.

Yan, Z. and Nie, Z. 2004. A fair comparison of the randomized response strategies. Unpublished.

Yates, F. and Grundy, P.M. 1953. Selection without replacement from within strata with probability proportional to size. *JRSS* B 15, 253–261.

Zacks, S. and Bolferine, H. 1992. *Prediction Theory for Finite Populations*. New York, NY: Springer–Verlag.

Zdep, S.M. and Rhodes, I.N. 1976. Making the RR technique work. *Public Opin. Quart.* 40, 531–537.

Zdep, S.M., Rhodes, I.N. Schwartz, R.M., and Kilkeun, M.J. 1979. The validity of the randomized response techniques. *Public Opin. Quart.* 43, 544–549.

Index